Management Systems for Construction

Alan Griffith
Paul Stephenson
Paul Watson

Longman

Pearson Education Limited
Edinburgh Gate
Harlow
Essex CM20 2JE, England

Published in the United States of America
by Pearson Education Inc. New York

Co-published with The Chartered Institute of Building through
Englemere Limited
The White House
Englemere, Kings Ride, Ascot
Berkshire SL5 8BJ, England

First published 2000

ISBN 0 582 31927-7

British Library Cataloguing-in-Publication Data

A catalogue record for this book is
available from the British Library.

Set by 35 in 9/13pt Palatino
Printed in Malaysia, VVP

Contents

List of figures   vi
List of tables   x
Preface   xii
Acknowledgments   xiv

1 Introduction   1
Preamble   1
Time planning and control   2
Financial planning and cost control   3
Quality and performance   3
Health and safety regulation and implementation   4
Environmental evaluation and control   5
Information technology and communication systems   7
The influence of management systems on organisation and
    human resources   8

2 Time planning and control systems   11
Introduction   11
The need for planning and control   11
The role of the planner   13
Stages of planning   14
Methods of programming   17
Monitoring and controlling progress   37
Managing resources   40
Time/cost optimisation   42
Multi-project planning   47
Computer software for planning and control   51
Summary   62
References   63

3 Financial planning and cost-control systems   64
Introduction   64
Marginal costing and break-even analysis   64
Cash flow analysis and budgetary control systems   75
Capital investment appraisal   88

Investment appraisal example 92
Conclusion 101
References 101

4 Quality systems and performance 102
Introduction 102
The value of quality assurance in construction 102
Quality assurance systems 103
BS/EN/ISO 9000 series quality assurance systems 105
Cost implications of the certification process 106
The constituent parts of BS/EN/ISO 9000 series quality systems 108
The facts about BS/EN/ISO 9000 series quality systems 108
Testing the theoretical advocated advantages of implementation 113
Problems associated with implementation 119
A brief comparison of modernist and postmodernist assumptions 120
Generic overview of staff project roles within a quality assurance
   system 126
Generic outline of a project quality assurance system 127
Total quality management 141
Summary 147
References 147

5 Health and safety regulation and implementation systems 149
Introduction 149
Health and safety law 151
The Construction (Design and Management) regulations 1994 157
Project health and safety documentation 167
Management of health and safety 169
The designer's health and safety management system 179
The principal contractor's health and safety management system 185
Health and safety management: example 206
References 251

6 Environmental evaluation and control systems 253
Introduction 253
Environmental management and construction 255
The need for environmental management and EMS 259
Environmental management systems and standards 261
Environmental management and the corporate organisation 265
Environmental management and the construction project
   organisation 279
Environmental management system development checklists and
   outline framework 294
References 298

7	**Information technology and communications systems**	300
	Introduction	300
	Information in the construction process	300
	Information technology in the construction industry	301
	Strategic issues of information technology	306
	Costs and benefits of information technology	308
	Analysing and developing information systems	309
	Communications and networks	315
	Implementing information systems in construction organisations	329
	Human–computer considerations	331
	Staff education and training	332
	Effective information systems management	333
	Future developments	335
	Summary	336
	References	337
8	**The influence of management systems on organisation and human resources**	340
	Introduction	340
	Management systems	341
	Influence of management systems on organisation	344
	Sub-systems as management support services	346
	Influence of management systems on human resources	347
	Commitment to management systems	348
	References	350
	Index	351

List of figures

2.1	Planning stages and types of programme	15
2.2	Matrix showing possible types of project and programme	16
2.3	Bar chart of earthwork operations	19
2.4	Bar chart of operations for excavation work	20
2.5	Bar chart for residential development	21
2.6	Bar chart of sub-operations for drainage work	22
2.7	Representation of activities in arrow diagram format	25
2.8	Representation in an arrow diagram of activities that can start at the same time	25
2.9	Dummy activities in an arrow diagram	26
2.10	Types of float associated with an activity	26
2.11	An example of an analysed arrow diagram for a small project	28
2.12	Types of logic relationship between activities in precedence diagram format	29
2.13	An example of an analysed precedence diagram for a small project	30
2.14	An example of the structure of an arrodence diagram	31
2.15	An example of an analysed arrodence diagram for a small project	31
2.16	A non-complex network of sixteen activities	32
2.17	A complex network of sixteen activities	32
2.18	An example of an analysed ladder construct	33
2.19	Network sections with simple and complex interfacing	34
2.20	An example of a sectionalised network for the construction of an industrial building complex	35
2.21	An example of sub-networks for part of a business centre development scheme	36
2.22	An example of progressing work on a bar chart	38
2.23	Part of an arrow diagram drawn to a time-scale	38
2.24	Progressing work and updating information directly on a network	39
2.25	Progressing activities on a precedence diagram	39
2.26	An example of a resource profile for activities	41
2.27	An example of resource smoothing by using float for non-critical activities	41
2.28	A linear relationship between time and cost for an activity	43
2.29	A multi-linear relationship between time and cost for an activity	43
2.30	A discrete time and cost relationship for an activity	44

2.31	A non-linear relationship between time and cost for an activity	44
2.32	Cost curves to establish the least cost for optimum project duration	45
2.33	Part of an analysed network for compression	46
2.34	An example of multi-project planning with complex interfacing	49
2.35	Multi-project planning using bar charts to illustrate links for work continuity	50
2.36	Network diagram of a building project	53
2.37	Entering project information for a new project	54
2.38	Defining calendars for a project	54
2.39	Entering activity details on the activity form	55
2.40	Establishing relationships between activities using the successor dialog box	56
2.41	Dialog box for entering constraints on an activity	56
2.42	Defining resources for a project	57
2.43	Allocating resources to activities	57
2.44	Resource profile for the number of labourers	58
2.45	Resource table for the cost of selected resources	58
2.46	Assigning target dates to activities	59
2.47	Updating details of an activity	60
2.48	Selecting activity box configuration	60
2.49	Selected display of activities with a cosmic view of a network	61
2.50	Selected section of network showing progress on activities	61
3.1	Segregation of semi-variable costs	66
3.2	Traditional break-even chart	68
3.3	Profit graph example	69
3.4	Pictorial representation of the concept of marginal costing	70
3.5	Solutions to the example strategies presented on a profit graph	72
3.6	Break-even chart example	73
3.7	Identification of average costs graph	74
3.8	The dynamic control feedback loop	75
3.9	School contract example contract programme	77
3.10	Standard 'S' curve graph for the school contract example	80
3.11	Cash flow graph for the school contract example	81
3.12	Cumulative cash flow graph example	82
3.13	Allocation of cumulative costs for the school contract example	83
3.14	Plotting the negative and positive cash flows for the school contract example	84
4.1	The route to BS/EN/ISO 9000 series certification	107
4.2	Contents of the quality and procedures manuals	109
4.3	Procedural pro forma front sheet	111
4.4	Pro forma procedure format (purpose)	112
4.5	Pro forma procedure format (scope and responsibility)	112
4.6	Pro forma procedure format (procedure)	113

4.7	Analysis of field research pertaining to theoretical advocated advantages of implementing BS/EN/ISO 9000 series	116
4.8	Noted disadvantages of implementing BS/EN/ISO 9000 series	118
4.9	Implementational change model	124
4.10	Project quality assurance documentation file front cover: example pro forma	130
4.11	Quality assurance documentation details cover sheet: example pro forma	131
4.12	Document issue: example pro forma	132
4.13	Drawing issue: example pro forma	133
4.14	Communications receipt and despatch: example pro forma	134
4.15	Information request: example pro forma	135
4.16	Receipt of information: example pro forma	136
4.17	Request for architect's instructions: example pro forma	137
4.18	Receipt of architect's instructions: example pro forma	138
4.19	Architect's verbal instructions: example pro forma	139
4.20	Internal audit for subcontractors: example pro forma	140
4.21	Interim audit of suppliers: example pro forma	141
4.22	Materials delivery: example pro forma	142
4.23	Weekly report: example pro forma	143
4.24	Generic model for the implementation of total quality management	146
5.1	Profile of fatal accidents in the construction industry, 1986–96	150
5.2	European Directives adopted under Article 118A of the Treaty of Rome	154
5.3	The Management of Health and Safety at Work Regulations 1992 (UK regulations implementing European Directives)	156
5.4	Application of the Construction (Design and Management) Regulations 1994	160
5.5	Management systems for project health and safety to accommodate the CDM Regulations	170
5.6	Inputs to the development of the pre-tender health and safety plan	174
5.7	Designer's health and safety management system during project evaluation and development (design review)	180
5.8	Designer's hazard identification record: suggested pro forma	182
5.9	Designer's risk assessment record: suggested pro forma	184
5.10	Principal contractor's health and safety management system	186
5.11	Principal contractor's management responsibilities: suggested pro forma	194
5.12	Principal contractor's document register: suggested pro forma	195
5.13	Principal contractor's risk assessment: suggested pro forma	196
5.14	Principal contractor's safety method statement review: suggested pro forma	197
5.15	Principal contractor's site safety co-ordination record: suggested pro forma	198

5.16	Principal contractor's permit to work: suggested pro forma	199–200
5.17	Principal contractor's safe working procedures: suggested pro forma	201
5.18	Principal contractor's safety incident investigation: suggested pro forma	202
5.19	Principal contractor's accident report: suggested pro forma	203
5.20	Principal contractor's site safety inspection: suggested pro forma	204
5.21	Principal contractor's induction and further training: suggested pro forma	205
6.1	Framework for project environmental management	260
6.2	Organisational development of an environmental policy	266
6.3	Key elements of a standard environmental management system	268
6.4	Contribution of knowledge to the final project brief	284
6.5	Environmental considerations in developing a project's final design	285
6.6	Functional areas of contract administration in developing a contractor's environmental management programme	291
6.7	Guidelines for the development of an environmental policy statement	296
6.8	Corporate system development	297
6.9	Project system development	298
7.1	Communications in relation to a company	302
7.2	Outline stages of the traditional 'ideal' construction process	304
7.3	Relationships between strategies	306
7.4	Stages of the systems development life cycle	311
7.5	Stages in a prototype systems development	312
7.6	Stages in the evaluation and selection of applications packages	313
7.7	Stages in end-user systems development	314
7.8	Stages in outsourcing systems development	314
7.9	Types of transmission between communicating computer systems	316
7.10	An estimating department local area network	318
7.11	A metropolitan area network for a construction company	319
7.12	A wide area network for a construction company	319
7.13	Connection of company networks to networks of external parties	320
7.14	Network typologies	321
7.15	A typical arrangement for a construction company's communication links	322
7.16	An example of client/server computing	322
7.17	An example of connecting company LANs to the Internet	323
7.18	A representation of a firewall between networks	328
7.19	Cumulative costs associated with an information system	334
8.1	An organisation viewed as sub-system elements supporting the system and its core business	343
8.2	Management stratification within the organisation of a sub-system	345

List of tables

2.1	An example of establishing work durations from a net cost estimate	24
2.2	Critical path method definitions and equations	27
2.3	Calculated times for the network shown in Figure 2.11	28
2.4	Cost information associated with the network shown in Figure 2.33	46
2.5	Summary of time and cost information for compressing the network shown in Figure 2.33	48
2.6	Estimated durations and resource allocation details for building project activities	53
3.1	School contract example cost/value data	76
3.2	Loading the school contract example with costs	77
3.3	Loading the school contract example with profits	78
3.4	The school contract example loaded with cost and profit elements	78
3.5	Tabulated sheet for the school contract example	79
3.6	Calculation of interest payments for the school contract example	83
3.7	Calculation of the funding required for the school contract example	84
3.8	Calculation of the average monetary sum lock-up for the school contract example	85
3.9	Breakdown of elemental costings for the school contract example	85
3.10	Percentage completions incorporated matrix for the school contract example	86
3.11	Variance analysis table for the school contract example	87
3.12	Individual variance calculations per cost centre matrix for the school contract example	87
3.13	Data for appraisal techniques example	89
3.14	Average annual percentage rate of return	90
3.15	Tabulated NPV data at 14%	90
3.16	Tabulated NPV data at 10% and 12%	91
3.17	Net present value tables for examples	93
3.18	Tabulated data for Project 'A': payback method	94
3.19	Tabulated data for Project 'B': payback method	94
3.20	Tabulated data for Project 'A': average annual percentage rate of return method	95
3.21	Tabulated data for Project 'B': average annual percentage rate of return method	96
3.22	Data for Project 'A': net present value method	96

3.23	Data for Project 'B': net present value method	97
3.24	Data for Project 'A': internal rate of return method	97
3.25	Data for Project 'B': internal rate of return method	98
3.26	Data for Project 'A': payback period including net present value	99
3.27	Data for Project 'B': payback period including net present value	100
3.28	Matrix analysis: comparing results for Project 'A' and Project 'B'	100
4.1	BS/EN/ISO 9000 series quality system elements	109
5.1	Evaluation criteria for severity of harm	171
5.2	Evaluation criteria for likelihood of occurrence	171
5.3	Determination of priority rating for risk	172
6.1	The environmental effects of construction	257
6.2	Checklist of factors to consider in determining the environmental position of the organisation	295
6.3	Checklist of key steps in developing the environmental approach	295
6.4	Checklist for considering the preparatory environmental review	295
6.5	Checklist for prioritising organisational activities and their aspirations	296

Preface

Management Systems for Construction focuses on the application by principal contracting organisations of six key management concepts when undertaking construction projects. The six concepts and the management functions, systems and procedures that they give rise to are essential to project success. These are the management of time, cost, quality, health and safety, environmental impacts, and information and communications. Why are these particular management concepts included? They are all essential management support services underpinning both the success of the construction project and the core business of the parent organisation. They must be addressed in almost all construction organisations and while undertaking almost all construction projects. They all use a systems approach which links the corporate organisation with the project-based organisation in a holistic way, and they are all essential to the organisation in meeting increasingly stringent legislation and regulation, changing technology, demanding client expectations, performance auditing and wider public accountability. Also, these concepts have not until now been brought together in a single book. It may be asked why this book has not included additional management functions which are also essential to the successful undertaking of any construction project, such as the management of plant and equipment, materials and manpower. That is simple: those aspects have been well covered in other books.

The purpose of this book is to provide an introduction to each of the six key management concepts. The focus is the application of each concept through the development of management systems and procedures by the principal, or main, contracting organisation. The primary aim is to inform construction professionals, academics and students of the current status of each management concept and how these concepts can be applied in the construction process. Each chapter can easily merit a dedicated book in its own right. In this book, each chapter represents an introduction to the topic, and the reader is directed to more detailed works and, in relation to particular concepts, authoritative references such as national and international industry sector-specific standards, regulations and guidelines.

While the focus of this book is on the management activities of the principal contracting organisation, important reference is made to the complementary roles, responsibilities and activities of other parties to the construction process. For example, to appreciate the principal contractor's construction-phase health and safety plan, a requirement of the Construction (Design and Management) Regulations 1994, it is also essential to appreciate the pre-tender health and safety plan compiled by the planning supervisor, aided by the design team. Similarly, to appreciate the implementation of environmental management systems by the principal contractor, it is helpful to be aware of complementary activities undertaken by the client and consultants during project evaluation and development.

Throughout the various chapters in this book reference is made to 'systems'. This is the term used most frequently to represent those organisational protocols which address the management of a particular concept in both organisational and project situations, for example a quality management system – 'the organisational structure, responsibilities, procedures, and resources needed to implement quality management' (Paradis *et al.*, 1996). Respect to this understood convention is paid in this book. Nevertheless, it is contended that the organisation itself can indeed be termed 'the system', around which management functions, for example quality, are sub-systems. This aspect is explored in the final chapter of this book.

This book does not intend to be prescriptive or claim to describe best practice. Individual construction organisations will find their own best way of developing their corporate and project systems given their particular circumstances and specific influences. What this book does is for each management concept present a potential approach to system development and application.

Reference

Paradis, G. W., Small, F. and Information Mapping Team, ISO (1996) *Demystifying ISO 9000*, Addison Wesley, Massachusetts, USA.

Acknowledgments

The authors acknowledge *all* those persons who have contributed to and supported the writing and publication of this book.

1 Introduction

Preamble

The high levels of construction management skills that are assembled and structured into effective project teams and the demanding responsibilities that are assumed for any construction project are principally a function of the degree of specialisation that has evolved within the construction industry. Specialisation is intrinsic to construction. Large and complex projects are characterised by their teams of specialists. Traditionally, these specialists have included planners, estimators, surveyors and construction managers with the focus on planning, monitoring and controlling project tasks and their sequence, resourcing, duration and cost. In recent times, these specialists have been joined by an additional host of managers, with the most prominent responsibilities focusing on quality, health and safety, environmental impact, and information technology.

Today's construction projects could not function without some degree of specialisation. Each specialist contributes to the project by following established professional working practices and by implementing particular systems, manifest in sets of procedures which translate each concept into management tasks. Specialists guide the various stages of the construction process, and the systems which they adopt assist them to maintain the many, varied and complex arrangements that need to be made. Furthermore, clients and the corporate management of those organisations involved with the construction process require an assurance that organisational and project procedures are being clearly determined and followed. This is absolutely essential in a construction environment, where there is increasingly stringent regulation, performance auditing and wide public accountability.

This book provides an introduction to six specialist and key management concepts and the management functions, systems and procedures that they give rise to. These are the management of time, cost, quality, health and safety, environmental impact, and information and communication. The organisational focus is the principal contracting organisation. All of these management support services underpin the success of both the construction project and the core business of the parent organisation. Within each chapter, the purpose is to provide an introduction to the management concept and the systems that may be considered in managing that aspect. The final chapter looks at the influence of those management concepts and systems on the organisation and on human resources.

As stated in the preface to this book, there is no intention to be prescriptive or to describe best practice. A contracting organisation must make up its own mind as to how to manage each key concept at both the corporate and construction project levels given its own particular circumstance.

Time planning and control

The management of time is a fundamental and predominant consideration for any construction project. Irrespective of the type of project, its size and resourcing there will an essential requirement to plan, monitor and control all activities against the project's duration. Within construction, time planning and control is more involved than simply viewing the planning, measuring and control discipline as a routine cycle. The unique nature of the construction process presents complexities, uncertainties and changing circumstances which must be accommodated within the planning and control system used.

Planning will be carried out at the pre-tender, pre-contract and contract stages of most projects, and each stage represents an important activity for a contracting organisation. As almost all but the smallest projects comprise a large number of interdependent items of work and involve many participants, reliable plans and accurate progress-recording mechanisms become all the more essential to project success. The organisation requires a sound time-planning and control system which allows not only efficient and effective management of an individual project but also the likely need to manage multiple projects simultaneously. Again, time-planning and control systems are far from being a simple management tool in application.

Unlike quality, safety and environmental management systems, which are characterised by formal systems meeting the requirements of national and international standards and certification schemes, time-planning and control systems, and for that matter financial-planning and cost-control systems, are systematised within the particular management methods and tools utilised. The management methods applicable are well recognised in sophisticated Gantt charts and the variety of network-based approaches available. The tools used really give the orientation of the management system, since the system is formed around the computing software implemented. Most organisations will have their own preferences for particular software applications for time planning and control. Once familiar and satisfactory to the organisation, it is perhaps one specific application or a small number which forms the hub of time planning and control at both the corporate and project levels.

Chapter 2 presents a detailed introduction to the concept, principles and practices of time planning and control. The need for planning and control and the role of the planner are explained through reviewing the various stages of planning. Methods, tools and applications follow which look at the individual project and multi-project planning and progressing techniques. While specific software applications are not recommended, the use of generic computer software packages around which the organisation may form its time planning and control management systems is explored. It is advocated in this book that a successful approach to time management will be based around a well-understood and effective computer-based project system.

Financial planning and cost control

The planning and control of a construction organisation's finances are crucial to its long-term well-being and survival. Therefore, it is vital that the senior management of an organisation fully appreciates that it must have a holistic perspective of its financial activities. The cumulative nature of cash flow is a vital aspect in the planning and control of the overall business and operational activities within an organisation.

Within the planning and control of operational processes, managers have to make decisions which have a strong bearing on financial issues at both the corporate and construction project levels. Various techniques are encapsulated within a financial-planning and cost-control system, and these can be set within a decision-making framework.

Though identified and explained in a linear fashion, these techniques are not mutually exclusive and should be viewed as a range of management control tools to be incorporated within a coherent financial control system. Only when this has been achieved will the holistic aspect of planning and control be fully achieved by the organisation.

Quality and performance

Quality has without doubt become a major competitive factor for construction organisations to consider. The proliferation of construction-related firms which are now certificated under the BS/EN/ISO 9000 series underpins this importance. Also, some construction organisations have already implemented or are considering total quality management (TQM) initiatives. Both strategies are being pursued as a means of gaining a competitive advantage. However, the critical issue is that for any organisation to succeed and prosper in the future any competitive advantage must have a high degree of sustainability. It is this issue of sustainability that provides the key to the implementational processes associated with quality systems.

Construction organisations should have a comprehensive understanding of what BS/EN/ISO 9000 quality systems or TQM is designed to achieve. For example, if the senior management of a construction organisation implements BS/EN/ISO 9001 expecting to obtain an immediate improvement in efficiency then it is unlikely to obtain it. This is due to the intrinsic nature of the quality system and what it seeks to achieve.

In Chapter 4, the advocated advantages of quality systems are tested and the main problematic issues of implementation are identified and examined. In addition, generic implementation models are suggested as outline information for any organisation embarking upon the implementation process. A quality system that is dynamic in nature will greatly assist in the attainment of a sustainable competitive edge, but a system should not be merely invoked but rather be founded upon an organisation's distinctive competence.

Health and safety regulation and implementation

The management of health and safety is unequivocally one of the most important functions of construction management. Construction work is intrinsically hazardous. Injuries to persons on and around construction sites occur regularly. It is perhaps fortuitous that many injuries are minor, but others are serious and some are fatal. The construction industry has over the last 20 years suffered a poor health and safety record. While the number of fatalities has shown a welcome decline in the 1990s, this should not encourage complacency. Construction management has a perpetual and unswerving challenge to ensure a safe working environment.

The Construction (Design and Management) Regulations 1994 introduced welcome and much-needed legislation to construction health and safety. The CDM Regulations are concerned with the management of health and safety throughout the whole construction process. Responsibility is clearly and specifically placed upon clients, designers and contractors to be proactive in the planning, co-ordination and management of health and safety. The Regulations focus on identifying the potential hazards to health and dangers to safety through each stage of the construction process, together with the assessment of their risk.

The CDM Regulations require a two-stage approach to health and safety planning and management. The first stage focuses on the project evaluation and development processes with the object of producing a pre-tender health and safety plan. The second stage focuses on the production site processes with the object of producing a construction health and safety plan. It is the essential element of planning within each stage which forms the basis for a systems management approach, within which risk assessment is the central theme.

Effective health and safety management systems and working procedures are the goal of the main parties to the construction project. The lead consultant, representing the client and working with sub-consultants, is charged with delivering a pre-tender health and safety plan and implementing management procedures that make a full contribution to project health and safety. The principal contractor is charged with delivering a construction health and safety plan. Moreover, the contractor must establish management systems and working procedures which ensure the maintenance of safe working conditions.

Well-formulated health and safety management systems will identify, assess and control risk both within and across the professional boundaries of the parties. Feedback loops within the designer's and the contractor's approach will ensure that information is not only directed within the span of control of the individual party but also contributes to the management processes within other systems. Within the context of the CDM Regulations, a specific outcome from the management approach to health and safety is the delivery of a health and safety file – a complete profile of health and safety planning and management throughout the construction project.

That a systematic approach to health and safety management in construction is essential is not in question. Government has recognised and the industry accepted that the undesirable accident record of construction must be improved. The CDM Regulations place clear and unambiguous responsibilities upon the main contracting parties to deliver health and safety management. It is suggested in this book that the implementation of a clearly conceived, formally structured and well-organised health and safety management system is the most appropriate way for participants to ensure that they make a full contribution to providing a safe construction process. Health and safety management systems (H&SMS) development is following in much the same way as quality management and environmental management. Moving towards accredited certification schemes, an H&SMS should meet the requirements of BS 8800, the UK's specification for health and safety management system development.

Environmental evaluation and control

The construction industry has a significant effect upon the environment whenever, wherever and in whatever form construction works are carried out. Almost all organisations within the construction industry face increasing pressures to broaden their understanding of environmental matters. Moreover, they must respond to commercial and public expectations for improved environmental business performance and commitment to environmental safeguards.

Awareness of environmental management within the industry is increasing. Influenced by the introduction of BS 7750 – 'Specification for Environmental

Management Systems' – in 1992 and its international counterpart ISO 14001 in 1994, environmental management systems, or EMS, are considered by many construction clients, consultants and contractors as a positive way forward in response to increasing environmental demands.

Environmental management standards recommend that an organisation should develop, implement and maintain a structured management system commensurate with the environmental policy, strategy, aims and objectives that it sets in the course of running its business. It must have in place formalised mechanisms for the evaluation and control of its environmental effects. Furthermore, it must ensure that the system meets all the current environmental legislation which regulates its business products and activities.

The contribution made to an organisation through the implementation of an environmental management system can be considerable. Benefits may extend well beyond satisfying particular environmental legislation. Benefits may be conceived both within the intra-organisational framework and operations and within the external business environment.

An environmental management system may take the form of a standard system which meets the specification of a recognised national or international system, or it can be a bespoke system to meet the individual needs of the organisation and its business. Systems do not need to be established in one fell swoop. They can be built up gradually as various sets of procedures are introduced and embedded within the culture and management practices of the organisation. Certainly, where environmental management systems have been established within construction organisations experiences have been positive, and such vision and commitment is seen as the most proactive way of meeting the environmental business challenges of the future.

Within the construction process, environmental management has made some strong inroads. In construction procurement, some clients are pre-qualifying prospective consultants and contractors based on their level of environmental proactivity. It is overwhelmingly likely that there will be increasing pressure upon consultants and contractors to support environmental management from public sector and major private sector clients. This will be particularly prominent for projects with significant environmental sensitivity. In time, it is possible that for an organisation to be placed on a tendering list it must validate its experience in the use of a formalised environmental management system.

The principal focus of environmental management system implementation is at the production site by the main contractor. Certainly, this is the most obvious and appropriate application of environmental management, since the production stage is where the greatest environmental effects are invariably manifest. Contracting organisations have a key responsibility within construction management to safeguard the environment in the course of carrying out their works. The systems and procedures that they adopt and the level of commitment that they demonstrate will be enhanced where support and commitment come from corporate management. For this reason, environmental management works most effectively where the parent company establishes a corporate environmental management system which is translated into a system of procedures implemented at the project site level. Such an approach is suggested in this book. Emphasis is placed on developing, implementing and monitoring this appropriate environmental management system.

In addition, the understanding and application of environmental management principles is important at the project evaluation stage. Here the process of environmental impact assessment is vital in identifying and evaluating the potential environmental effects of the project. Also essential to a successful project outcome is environmental evaluation in the briefing and design stages. These stages are crucial as they provide invaluable information to the client and designer throughout the project evaluation and development process. In this book, these aspects are also addressed and again it is suggested that a systematic approach to their undertaking is essential if the project is to reach a successful outcome.

Spiralling demands for more environmentally empathetic evaluation, design and construction, and in particular increasingly stringent regulation, mean that environmental management in all its forms and at all stages of the construction process will become more prominent in the future. If one doubts such an assertion, one has only to reflect on the evolution of quality management within construction, an aspect mirrored closely by the conceptual development of environmental management.

Information Technology and Communication Systems

The management of information and communication is increasingly important to contracting organisations. Moreover, an understanding of the technology available for making information and communication more efficient and effective influences organisational success. The construction process is characterised by the reliance on timely and appropriate contractual and technical information. Information and communication are complex within construction. In view of the many participants, the different levels of corporate and project based organisations and management, it is apparent how the human and technical barriers to the flow of communication and information materialise.

The quantity of information generated within the construction processes is immense. The speed and simplicity with which communication and information can be created and erased is staggering. This is compounded when electronic communication and information management systems are introduced by organisations and as technology rapidly advances. The security of communication and information is now an issue. It is essential that corporate and project information, both within the contracting organisation and between the parties to the construction processes, is reliable and safe.

In many applications the information, or data, available may be for reference and therefore unchanging, but, such is the nature of construction, much of the data used is for operational purposes and will require updating, so constant change to information is inevitable. Given the ever-evolving nature of construction projects and the variation to originally envisaged works the currency of information must be ensured.

A further dimension to communication and information within a contracting organisation is the interaction between corporate administrative offices: the head; regional; and area; and the many individual, and possibly widely dispersed, construction project sites. Available technology allows endless possibilities for communication and information management between users at these locations by utilising computer networks.

Information technology and communication systems established within a contracting organisation are defined by the specific computer systems adopted and the software used. System establishment will be based on strategic consideration given at corporate level to whole organisation needs for communication and information management down to operational needs at the project site. While an organisation will devise its own approach from particular needs and circumstances, there are many generic aspects which can be explored.

This chapter does not specify any particular communication and information technology software but rather focuses upon these generic aspects. It starts with an overview of information within the construction processes and the need for information within construction organisations. Considerations for the development of information systems and communication networks follow, with insight into individual and group based support systems. Implementing information systems in a contracting organisation is explored together with the key considerations for hardware and software selection. The effect of communication and information management upon company employees is a paramount consideration and, therefore, the chapter examines the human-computer interface and the need for staff development, training and support. The chapter concludes with an overview of effective information systems management, the achievement of systems value and likely developments in information technology and communication systems.

The influence of management systems on organisation and human resources

The business environment today is demanding and always changing. Increasing attention is being paid to organisational performance and to the added value obtained from services and products. Organisations need effective strategic, directive and operational management supported by committed teams of employees if such demands are to be met. Many organisations have reoriented themselves from being morphostatic to being morphogenic in nature. Teamwork has taken on a new meaning as managers have assumed direct responsibility for running parts of the business. Construction organisations have not been exempt from such change. Lean organisations with their flat management structures and reduced workforces typify recent changes in the industry.

Management systems have developed along with structural change within many organisations. Management systems develop protocols and sets of procedures which provide structure and organisation. Where management systems have developed around a holistic and morphogenic vision then the potential to meet many organisational needs exists. Where systems are created without reference to the organisation as a whole or its other parts then there can be a potential for chaos.

The organisation and management of construction lends itself to a systems approach. As presented in this book, management systems are well established within the construction process, for example time, cost and quality management approaches. These have been joined in recent years by management systems for health and safety implementation, environmental evaluation and control, and information technology. All of these have embraced management concepts within sets of procedures which can be applied to the organisation in a systematic and coherent way.

One of the major issues in developing such systems is how the systems are perceived. In fact, the system is not the environmental management system or the quality management system. The system is the organisation. The management concepts which are embraced within the procedures are sub-systems which exist to serve the organisation and its core business. Therefore, environmental management, quality management and all other sub-systems are really support services. These services function for both the construction project organisation and the corporate organisation and add value to the total business.

In many construction organisations, in particular the larger ones, two tiers of management within each sub-system are likely to develop. One will deliver strategic and directive management at the corporate level, for example at head office and regional office level, while the other will deliver operational management, specifically at the project site. Such management arrangements have a vital influence on the structure of the organisation and also on the human resources deployed. The final chapter of this book looks at how management concepts are embedded within management sub-systems and how these develop and interrelate to support the business of the organisation.

It is not intended within the scope of this book to delve into the concept of the integrated management system (IMS). It is well recognised, however, that a logical extension of drawing construction management systems together under a parent system umbrella is to consider their configuration in an IMS format. Such is the current status of the management systems explored in this book that IMS development is some time away and that systems such as quality management, health and safety management, and environmental management, although obviously interrelated, are recognised separately for certification purposes. Notwithstanding this, it is apparent that much is being done by accreditation bodies to draw management systems together such that in the future construction organisations will be able to have an integrated system certified to international standards and accepted throughout the industry.

2 Time planning and control systems

Introduction

This chapter looks at the issues related to time planning and control. Basic principles are covered and also the techniques used in practice. These techniques are also extended to cover more complex issues related to projects, and practical examples are provided, where appropriate, to illustrate the application of theoretical concepts. In particular, the stages of the planning are considered, together with the role of the planner at the various stages of the construction process. The practicalities of planning time are explained through commonly used programming techniques, which also include monitoring and progressing of construction work.

Coupled with the planning of time is the issue of managing resources during the execution of work operations. This not only has an impact on the duration of individual operations or activities but also influences cost. Time/cost optimisation is therefore an important consideration for contracting organisations, and a technique used in the process is demonstrated to reduce project duration in a cost-effective way.

More involved planning issues are considered in the section covering multi-project planning. This is often a common occurrence for contractors, where work continuity and the effective utilisation of resources are important issues.

Finally, the chapter looks at computer application software used in the planning process. This includes identification of several automated features which provide planners with powerful tools with which to plan projects and produce information to assist in the decision-making process.

The need for planning and control

The planning and control of construction projects is now considered to be an essential requirement. Previous attitudes to the necessity and extent of planning required have changed over the years. Contractors themselves are establishing planning systems or, more commonly, are requested to do so by other parties to a contract.

Planning itself initially tends to focus the mind on the completion time for projects, and this is of interest not only to clients but also to contractors. Clients have vested interests in the handover and commissioning of a project. This may have commercial significance related to income generation through retail outlets, office development or manufacturing. Whatever the purpose of the built product, late completions will result in loss of income, which is often reflected in the amount of contractual liquidated and ascertained damages.

Contractors also have an interest in completing on time owing to the use of expensive resources and other work commitments. Late completions not only attract financial penalties but will also have an impact on project overheads, turnover and ultimate profitability.

The need for planning and control is also becoming more formalised as a requirement, depending upon the particular form of contract used (Cooke and Williams, 1998). While construction programmes are not usually contract documents, some forms of contract may request the contractor to produce a programme. In some instances, this may go further where the contractor is required to use a specified proprietary computer system to produce programmes as a requirement to tender for the work (Lester, 1991). Some forms of contract are worded in such a way that the client, or client's representative, must be able to ensure that the works are progressing satisfactorily. The preparation of a programme should therefore satisfy the client and also be of use to the contractor during the construction of the works.

The need to plan a project as early as possible is paramount to success. The programme can act as an ideal communication medium to illustrate logical thinking and the proposed sequencing of work. Parties involved in the project can view work sequences for evaluation, which may in turn provide the opportunity for new ideas and changes (O'Brien, 1993; Reiss, 1996).

Invariably, this may be of direct benefit to the contractor, since the identified sequencing of work may suggest alternative ways of constructing and economising on resources and subcontract prices. Ultimately, this may result in reduced project completion times and possibly reduced tender bids, making contractors more competitive (O'Brien, 1993).

As projects increase in complexity, installations and sophisticated systems become integral parts of designed projects. This adds to the necessity for planning. Complexity itself often suggests the involvement of specialist subcontractors, and programmes prepared at an early stage indicating the work requirements can assist in avoiding communication breakdowns. Technical complexity may also impact upon financial outlay for the client and have a direct bearing on risk. Effective planning will therefore indicate the requirement to construct and control work operations in an economic way (Ahuja *et al.*, 1994).

However, while plans provide a documented path for completion of a project, planning for planning's sake is of no use without committed implementation and use. Parties who have provided input to the planning process will be more likely to accept the plans produced and be committed to achieving the targets set (Ahuja *et al.*, 1994). Comparison of measured work against targets will also provide the opportunity for modifications and remedial action as appropriate. Updating programmes will bring home clearly the passage of time that has occurred, and work that should have been completed during a certain period (O'Brien, 1993). Programmes can also be evaluated and necessary changes identified for future work requirements. Accurate recording of progress and feedback on practical issues, buildability, and resource requirements and use will also provide valuable information for future estimating and tendering.

The role of the planner

The role of the planner is one that will vary depending upon the type of work and the contracting organisation. In small companies, the planner may be one individual who deals with other functions such as estimating and surveying. Larger companies may operate in a more structured way and have a planning department with personnel who specialise in specific stages of the planning process.

Inevitably, the planner's role is an important one. A planner may work as part of an estimating team with a specific involvement related to pre-tender planning. This may well entail information collection, assessment of construction methods, identification of cost-significant items of work and resource requirements. This will also include the preparation of overall programmes and liaison with other team members in support of tender preparation.

Additionally, planners may be involved with contract planning, possibly based at head office in the planning department, with responsibility for the planning of several projects. This may also include more detailed planning duties during the construction stages of a project.

While periodic site visits by a planner may be appropriate on some projects, other projects may be sufficiently large or complex to have a planner resident on site. During the construction phases, more detailed consideration of the contract programme will be required and will involve the preparation of short-term or stage programmes related to work requirements. Such plans will also need to be communicated to relevant parties. Additionally, progressing and updating programmes, scheduling of operations to support site staff, providing feedback and liaising with the site manager will be integral parts of the planner's responsibility. The role of the planner in documenting, providing and communicating information will therefore contribute to the overall success of a project.

Stages of planning

The extent of planning on construction projects tends to vary and is influenced by the type, size and complexity of the work. Construction companies will also deal with planning in different ways, which may be influenced by company policy and procedures.

Invariably, planning itself, in whatever amount of detail, tends to fall into three distinct stages:

- pre-tender planning
- pre-contract planning
- contract planning.

Pre-tender planning   Pre-tender planning during the tender period is closely linked to the estimating process and tender adjudication prior to submission of the bid. It is concerned with the outline details of the project in terms of methods of construction, resource identification and the establishment of a project duration. During this stage, the planner should work closely with the estimator and programmes produced should be consistent with methods, resources and outputs used in the estimate. Commercial considerations to establish a competitive bid should be dealt with at a later stage.

Pre-contract planning   On award of the contract, more detailed information requirements and organisational issues will be pursued as part of pre-contract planning before commencement on site. Invariably, the extent of this planning is not always sufficient, owing to the required start of the works by the client. Contractors would argue that more time availability during this stage would help in overall management of the contract, but the required completion date by clients may well hinder this.

During the pre-contract stage, further work can be done with greater emphasis on information collection, the organisation of resources, and compliance with contract conditions and regulations. Consideration can also be given to detailed phases and project completion times.

Contract planning   The contract planning stage which follows is concerned more with monitoring and controlling progress of the works. Target dates established previously can be used to indicate variance in work carried out, providing the opportunity for changes to be made and remedial action to be taken as required. Programmes will therefore be produced for each of the stages to suit the requirements of the project.

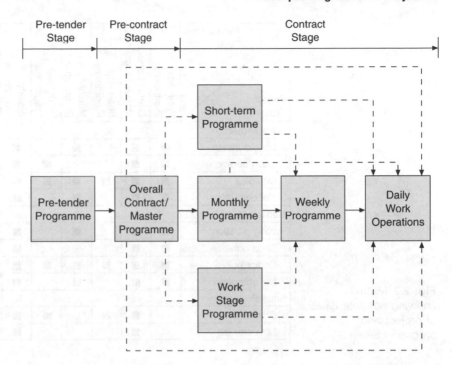

Fig. 2.1 Planning stages and types of programme

Types of programme

The question often arises as to what amount of detail is required when programming, and indeed, when is programming itself required. The answer usually lies in asking questions such as 'What is the programme required for?' 'Who requires it?' 'When is it required?' Senior management may require outline details only, whereas middle management may require a more detailed breakdown of work, with operational management requiring details of activities on a day-to-day basis.

Figure 2.1 shows a sequence of programming and suggests certain defined types of programming starting with a pre-tender programme, overall or master programme, and with progressive breakdown of detail to obtain short-term or stage programmes, together with weekly programmes and daily work operations. This does not mean to say that all these programmes are required on every project. There is no strict rule governing the extent of programming, which can vary considerably from one project to another. Figure 2.2 shows a matrix of possible programming requirements for different types of project.

Pre-tender programme

The first consideration of programming is carried out at the pre-tender stage. Usually, this is prepared along with a method statement to arrive at some overall plan of execution. The extent of detail at this stage can be considered as minimal, primarily to see if the work can be done in the time stated in the contract documents. Furthermore, the people most likely to be concerned with planning at this stage are senior management, so the overall contract in general

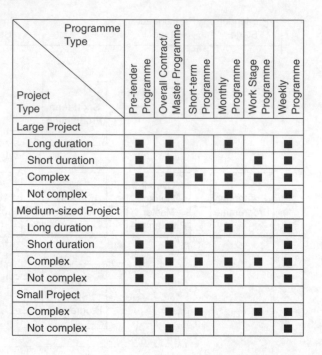

Project Type \ Programme Type	Pre-tender Programme	Overall Contract/ Master Programme	Short-term Programme	Monthly Programme	Work Stage Programme	Weekly Programme
Large Project						
Long duration	■	■		■		■
Short duration	■	■			■	■
Complex	■	■	■	■	■	■
Not complex	■	■		■		■
Medium-sized Project						
Long duration	■	■		■		■
Short duration	■	■				■
Complex	■	■	■	■	■	■
Not complex	■	■		■		■
Small Project						
Complex		■	■		■	■
Not complex		■				■

Fig. 2.2 Matrix showing possible types of project and programme

terms of work content is probably the only detail required. Primarily, the major or most important aspect at this stage is cost, so normally the programme should be in its simplest form, allowing judgments to be made. The pre-tender programme is prepared primarily to assist the contractor's bid, and the major cost elements may be the only items that are considered. The 80/20 rule in terms of the value of the work is of great importance when preparing an estimate, since major items are likely to influence a successful tender. Similarly, this same principle may apply to programming, and the major items or operations are therefore predominantly important. Consequently, the extent of pre-tender planning is totally dependent on the type of work. Tender periods may span only several weeks, so the amount of preparation that can be done in this period is often restricted.

Contract programme

The overall or master programme is often considered to be an update of the pre-tender programme. However, this is not always the case, since each programme serves a different purpose. The pre-tender programme acts as an aid to the contractor's bid, whereas the purpose of contract programming is to show in a realistic way how the timing of work is to be carried out. This does not mean to say that the data used for pre-tender assessment are not useful; it simply means that greater emphasis is placed on the work operations themselves.

It is not always necessary to adopt the proposed methods on site that were used at the time of tendering, since it is the responsibility of site management to determine the most appropriate methods of construction within estimating allowances. Admittedly, the estimator when preparing the estimate may not have chosen the best methods, but to ignore this information completely could prove financially disastrous. Generally, in large organisations the estimator does not price work in isolation, and the estimate and tender figure are usually the result of a team effort.

The important point is that this information should be communicated so that decisions can be made and a realistic programme produced. On large projects, the amount of detail required at this stage is restricted to the major phases of work to act as a guide to the whole contract period.

Short-term/stage programmes

The principle behind short-term or stage programming is to allow the overall programme to be considered in greater detail. The time period covering the work may be six weeks, possibly re-programmed monthly to correspond with site meetings and to allow for variations that may occur. On large projects, the programme may also cover particular stages of work to be executed over a short period of time, particularly where incentive schemes are in operation. The programme acts as a sub-programme, thus keeping the master programme up to date.

Weekly programmes and daily work operations

The weekly programme could be described as the lowest level of programming, indicating work to be carried out on a daily basis. The precise details of operations should be clear at this stage, together with resources required, and this also acts as a guide to maintaining the short-term programme. Daily work operations can be determined and linked to incentive schemes if applicable.

While the previously described types of programming are well known, it is not always necessary to carry them out in practice. As mentioned previously, many variables and conditions can occur that ultimately influence programming requirements.

Methods of programming

While the planning of construction projects has become well established in recent years, the extent of programming on contracts often varies widely. Some contractors produce overall programmes, while others consider that the most detailed analysis is essential to the success of a contract.

Various techniques have emerged over the years, so it is often confusing, or at least difficult, to decide which method to use. Current techniques would include:

- Bar charts
  - logic-linked bar charts
  - network-based bar charts
  - logic-linked network-based bar charts
- Network analysis
- Critical path analysis (activity-on-arrow)
- Precedence diagram (activity-on-node)
- Modified networks
  - PERT
  - GERT
- Graphical techniques
- Histograms and cumulative density curves
- Graphical analysis (network drawn as bar chart)
- Elemental trend analysis (line of balance)
- Vertical path method
- Mathematical models.

Slight variations in method do occur depending on the particular country where it is being applied or specific company which is using it, but generally all can be classified as one of the above. The problem lies in which method to adopt for a particular project. Clearly, certain methods are suited to particular types of work, but quite often the answer is not so clear-cut.

Several definitions and guidelines for the preparation of programmes have been prepared in the past covering the principles of programming and the information they should provide. However, it is difficult to predict every eventuality, since numerous variables may occur in practice and have a direct or indirect bearing on the eventual method chosen. Some of the main variables would include programme use, type of work, size, complexity, specialisation, duration, type of contract, conditions of contract, professional parties, management structure and resources.

From the various methods mentioned earlier, two methods are most predominant in practice, bar charts and network analysis, including both arrow and precedence conventions. Generally, companies tend to make selective use of network analysis, where complexity is a governing factor. Network analysis is generally employed for overall programming and bar charts for site use, the superior form being the logic-linked network-based bar chart. Bar charts, on the whole, seem to be the most common form of programming. They are quickly prepared, less involved and generally easily understood. Network analysis, however, has several advantages over the bar chart. The interrelationship of activities can be established, from which the various starting and finishing times can be obtained, together with the types of float for non-critical activities. Unfortunately, in practice this method has often met with dissatisfaction, since the diagrams produced can often appear quite complex and difficult to understand. Often this is not due to the programming

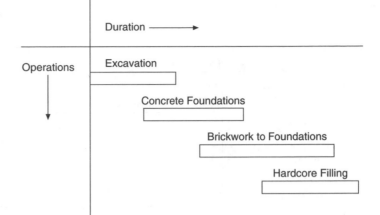

Duration ⟶

Operations

Excavation

Concrete Foundations

Brickwork to Foundations

Hardcore Filling

Fig. 2.3 Bar chart of earthwork operations

method itself but to inadequate training and insufficient appreciation of the technique.

It is not intended to cover all the various methods of programming in this section. The advantages and disadvantages of programming methods are published in numerous texts, some of which are referenced at the end of this chapter. Bar charts and network analysis, as the methods in most common use, are covered in detail below. It must be remembered, however, that programming is open to interpretation and quite often more than one method could be used to programme the same project. What is important, irrespective of method, is the usefulness of the programme as a communication document. The inputs and outputs required will vary depending upon what the programme is to be used for, and this in turn will be governed by the level and stage of usage.

Bar charts Bar charts probably represent the longest-serving form of scheduling available. Certainly, they are the most commonly used, and many contractors use them for all stages of programming, including large and small projects. A major advantage of the bar chart is that it is simple to prepare. Operations are represented vertically, with a bar representing the operation drawn on a horizontal time-scale indicating start and finish times and where the length of the bar represents the duration of the operation. This is illustrated in Figure 2.3. The final chart provides a graphic representation of the individual operations within the project.

Such charts are relatively easy to read and update and are readily understood by all levels of management, including site operatives. Thus they represent a useful communication tool. They can be useful to indicate to senior management the overall summary of major phases of a project while also being used to represent a small number of operations for short-term planning

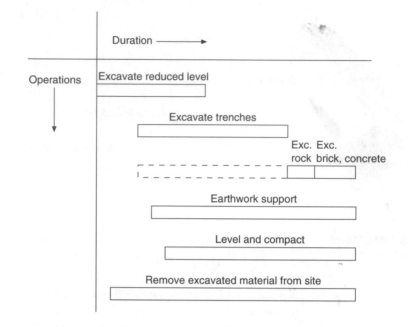

Fig. 2.4 Bar chart of operations for excavation work

over several weeks. Figure 2.4 shows a number of operations related to excavation work.

Probably the major disadvantage with bar charts is their inability to show sufficient detail. For simple projects this may not be a problem with a small number of operations, but with large, complex projects this may be problematic since operations need to be related through dependencies and constraints. While the overlapping of operations is shown on the chart, the logical sequence of operations is not. This leads to further problems of establishing the influence of changes to one or more operations on the overall programme, which may present difficulties in controlling operations and identifying potential problems. Figure 2.5 shows a residential development scheme represented in typical bar chart format with operations and estimated durations. Figure 2.6 shows a related bar chart of sub-operations for drainage work on the same project.

Network analysis

One method of overcoming many of the shortfalls or limitations in bar charts is by representing work in the form of a network. The network represents a sequence of activities related to the practicalities of construction. A network has obvious advantages over the bar chart in that the interrelationships of activities can be determined, from which a critical path of work sequence can be established. The various starting and finishing times of activities can also be obtained, together with the various types of float for non-critical activities.

In practice, this method sometimes creates dissatisfaction, since the diagrams produced can appear complex and difficult to understand, but appropriate training and application should help to overcome these problems.

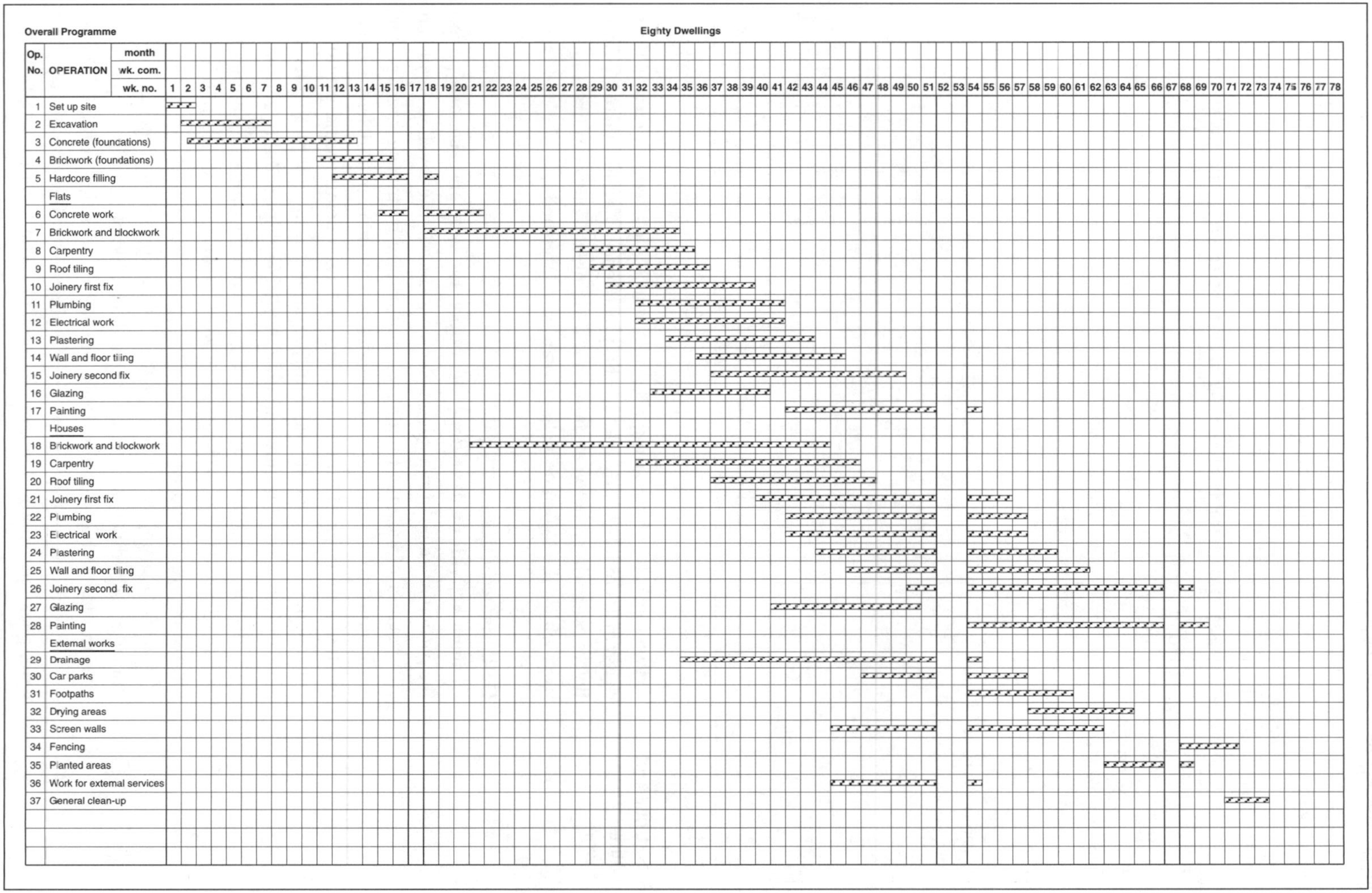
Overall Programme
Eighty Dwellings
Op. No.
OPERATION
month
wk. com.
wk. no.
1 2 3 4 5 6 7 8 9 10 11 12 13 14 15 16 17 18 19 20 21 22 23 24 25 26 27 28 29 30 31 32 33 34 35 36 37 38 39 40 41 42 43 44 45 46 47 48 49 50 51 52 53 54 55 56 57 58 59 60 61 62 63 64 65 66 67 68 69 70 71 72 73 74 75 76 77 78
1 Set up site
2 Excavation
3 Concrete (foundations)
4 Brickwork (foundations)
5 Hardcore filling
Flats
6 Concrete work
7 Brickwork and blockwork
8 Carpentry
9 Roof tiling
10 Joinery first fix
11 Plumbing
12 Electrical work
13 Plastering
14 Wall and floor tiling
15 Joinery second fix
16 Glazing
17 Painting
Houses
18 Brickwork and blockwork
19 Carpentry
20 Roof tiling
21 Joinery first fix
22 Plumbing
23 Electrical work
24 Plastering
25 Wall and floor tiling
26 Joinery second fix
27 Glazing
28 Painting
External works
29 Drainage
30 Car parks
31 Footpaths
32 Drying areas
33 Screen walls
34 Fencing
35 Planted areas
36 Work for external services
37 General clean-up

Fig. 2.5 Bar chart for residential development

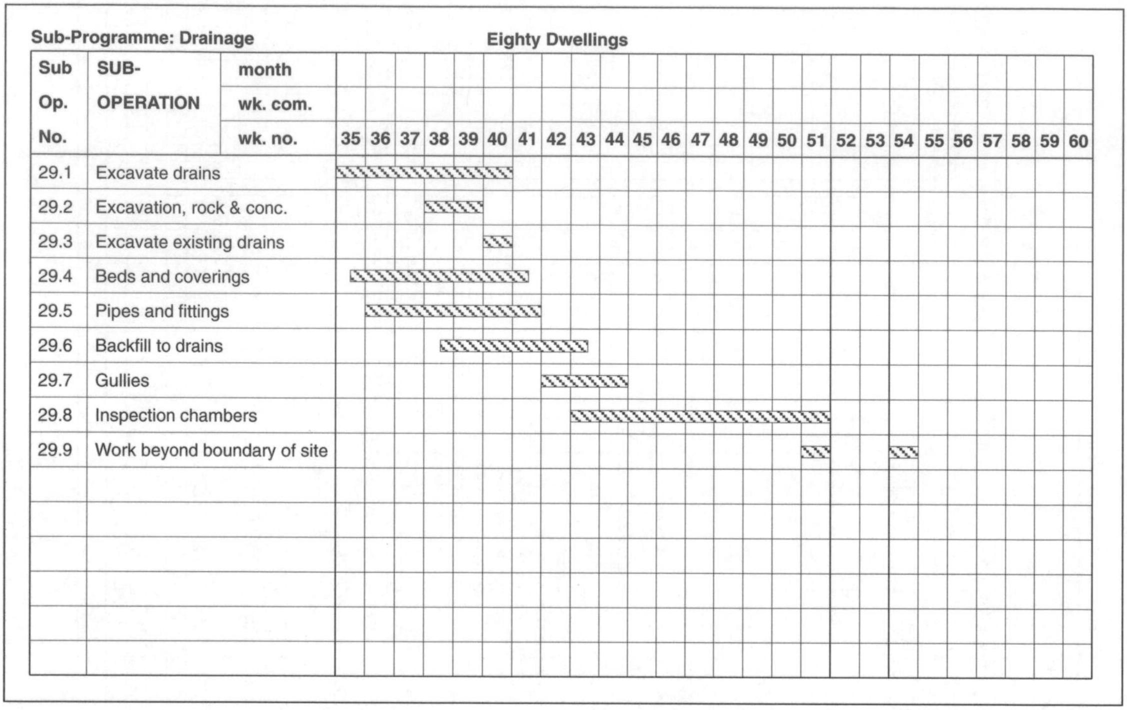

Sub Op. No.	SUB-OPERATION	month																									
		wk. com.																									
		wk. no.	35	36	37	38	39	40	41	42	43	44	45	46	47	48	49	50	51	52	53	54	55	56	57	58	59	60

Sub-Programme: Drainage   Eighty Dwellings

Sub Op. No.	SUB-OPERATION	wk. no.	35	36	37	38	39	40	41	42	43	44	45	46	47	48	49	50	51	52	53	54	55	56	57	58	59	60
29.1	Excavate drains		▨	▨	▨	▨	▨																					
29.2	Excavation, rock & conc.					▨	▨																					
29.3	Excavate existing drains						▨																					
29.4	Beds and coverings		▨	▨	▨	▨	▨																					
29.5	Pipes and fittings			▨	▨	▨	▨	▨																				
29.6	Backfill to drains					▨	▨	▨	▨																			
29.7	Gullies								▨	▨																		
29.8	Inspection chambers										▨	▨	▨	▨	▨	▨	▨	▨										
29.9	Work beyond boundary of site																	▨		▨								

Fig. 2.6 Bar chart of sub-operations for drainage work

Two methods of preparing networks tend to be used in practice: the arrow diagram (activity-on-arrow) and the precedence diagram (activity-on-node). Before preparing a network, however, defining activities is a first requirement from which the logic and network can be established.

Defining activities

Defining activities will be dependent upon the type and size of a project and the particular stage of planning being considered. The level of detail is normally a matter of judgment but may be influenced by management preferences or company policy and particular modes of operation. Insufficient detail would result in a network of limited use, and too much detail would produce a network which would be unnecessarily complex. An activity should therefore have a predictable duration and an identifiable responsibility and be of sufficient duration for remedial action to be taken if necessary. Activities may also have to relate to some predetermined work breakdown structure or cost centre for cost-control purposes.

A useful starting point is inspection of the tender documents. Analysis of drawings, schedules, bills of quantities and a list of significant quantities will give an idea of the scope of work. Further information collection by way of site visits may also reveal important findings. Identification of major items or work packages should therefore be possible. The level of detail can then be established and inclusion of other activities determined progressively.

Estimating the duration of activities

The duration of activities will largely be dependent upon the level of detail under consideration. This may be hours, days or weeks. The important point is that consistency must be maintained on the unit used. Whatever this is, the duration should be realistic and relate to the proposed method of construction using appropriate resources.

Activity durations are dependent upon the type and size of project, local conditions and other factors that are likely to influence productivity. They can be obtained from various sources, including outputs used in the estimate, times previously established through work study, past project data, or the planner's previous experience of similar work.

Table 2.1 shows an example of durations established from a net cost estimate based on a bill of quantities. Where project planning is a team effort, the chosen duration may well result from liaison between the planner, estimator and staff involved in the production process.

Suppliers and subcontract work

The type and size of project under construction will dictate the specific categories of work. Invariably, with the complexity of projects this will include some dependence upon suppliers and subcontractors. The extent of subcontract work will be determined by the main contractor in relation to work which is normally carried out by direct labour. The acquisition of materials and components, however, may have to go through pre-approvals in addition to purchase and delivery. Subcontract work also has to be discussed and agreed. These specific activities must therefore be defined as soon as possible for inclusion on the network.

Subcontracted activities may be representative of several trades, and it is important that all work is clearly shown. Subcontractors should therefore be asked to provide realistic durations for their work. Some activities may well be critical on a network, and discussions with suppliers and subcontractors will help to alleviate any potential problems. The network, in this sense, provides a useful communication medium for all parties concerned.

Basic network principles

One widely used method of producing networks is the critical path method (CPM). The network gives a graphical representation of the project and provides a sequence of activities showing interrelationships and interdependencies. From this logical planning phase, the scheduling of activities can then be considered to determine early and late starting and finishing times in order to establish the project duration. In doing so, the critical activities can be identified, together with non-critical activities and the availability of float. This also provides for further consideration of resource requirements throughout the duration of the project. Particular techniques of representing networks include arrow diagrams and precedence diagrams.

Table 2.1 An example of establishing work durations from a net cost estimate

Operation/ sub-operation No.	Operation/ sub-operation	Bill item	Item description	Quantity	Allocated labour and plant from estimate	Time allocated from estimate (hours)	Time allocated for sub-operation (hours)	Labour and plant/gang to be used	Time required for operation/ sub-operation (days)	Remarks
2.1	excavation: reduced level	85a	excavate reduced level < 0.25 m	56 m³	excavator + driver + banksman					
		85b	excavate reduced level < 1.00 m	45 m³	excavator + driver + banksman	18.48	26.13	excavator + driver + banksman	4	
2.2	excavation: trenches	85c	excavate trenches < 0.25 m	119 m³	excavator + driver + banksman	79.73				trench exc. includes all depths
		85d	excavate trenches < 1.00 m	248 m³	excavator + driver + banksman	49.60	129.33	excavator + driver + banksman	17	
2.3	excavate rock	85e	excavate rock in trenches	74 m³	excavator + driver + banksman + 2 labourers	29.60	29.60	excavator + driver + banksman + 2 labourers	4	
2.4	excavate brick and concrete	85f	excavate brick, conc. in trenches	93 m³	excavator + driver + banksman + 2 labourers	37.20	37.20	excavator + driver + banksman + 2 labourers	10	
2.5	earthwork support	85g	earthwork support	900 m²	labourer	198.00	198.00	2 labourers	25	
2.6	level and compact excavation	85h	level and compact excavation	14,511 m²	labourer	241.85	241.85	2 labourers + compactor	31	
2.7	remove excavated material	86a	remove excavated material from site	450 m³	lorry + driver	54.00	54.00	lorry + driver	7	

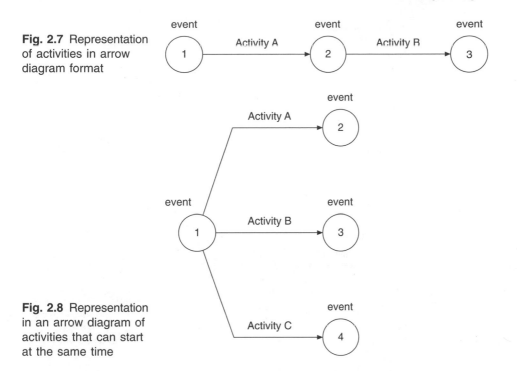

Fig. 2.7 Representation of activities in arrow diagram format

Fig. 2.8 Representation in an arrow diagram of activities that can start at the same time

Arrow diagrams

Arrow diagrams represent networks by the use of arrows and events. The arrow, representing the activity, is usually drawn from left to right, with events drawn as circles at the beginning and end of the arrow. The arrowhead represents the end of the activity, but the length of the arrow does not relate to its duration. The events segregate the arrows but themselves have no duration.

Activities are drawn in a sequential manner representing the order or logic of construction, as shown in Figure 2.7. From the defined activities, the order of representation is determined by deciding what activity can be completed before another can begin, and what activities can start once this activity has been completed. Assessment should also be made of activities that can start at the same time, as shown in Figure 2.8. It is important to maintain the logic of the diagram and to ensure that each activity has a unique event reference. This is achieved by the use of dummy activities, represented by dotted arrows. Dummy activities normally have no duration but are essential to indicate dependencies and restraints in order to preserve the logic of the network. This is shown in Figure 2.9.

Once the arrow diagram has been established and the logic confirmed, the network can be analysed and time calculations carried out by forward and backward passes. In the forward pass, the durations of activities are added together to establish the earliest start times and the longest path through the network.

Having established the project duration at the last event, the backward pass is carried out to establish the latest time that each activity can finish, again

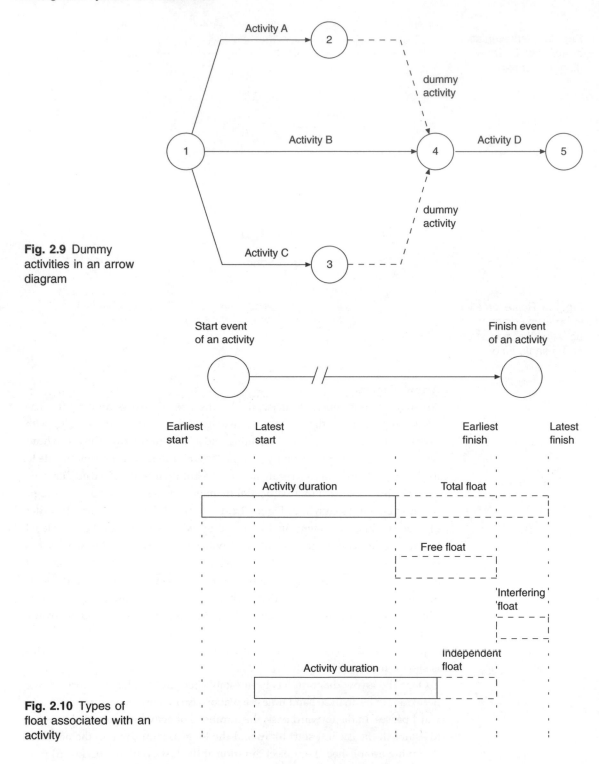

Fig. 2.9 Dummy activities in an arrow diagram

Fig. 2.10 Types of float associated with an activity

Table 2.2 Critical path method definitions and equations

Definitions

Duration (D)	Estimated duration of an activity
Earliest start (ES)	Earliest time an activity can start
Earliest finish (EF)	Earliest time an activity can finish
Latest start (LS)	Latest time an activity can start
Latest finish (LF)	Latest time an activity can finish
Forward pass (FP)	Calculation process through a network from start to finish to establish the earliest start and finish times for activities and project duration
Backward pass (BP)	Calculation process through a network from finish to start to establish latest start and finish times for activities
Critical path (CP)	A continuous sequence of activities representing the longest path through a network (more than one critical path may exist in a network)
Total float (TT)	Total amount of time that an activity can be delayed
Free float (FF)	Amount of time an activity can be delayed without delaying the early start of succeeding activities
Interfering float (INT.F)	Amount of time an activity can be delayed, but when used will affect succeeding activities
Independent float (IND.F)	Amount of time an activity can be delayed without affecting succeeding activities and which is not affected by preceding activities

CPM equations

Forward pass (FP)

ES	Project start for beginning activities
EF*	$= ES + D$

Backward pass (BP)

LS*	$= LF - D$
LF	Project finish for ending activities
For critical path (CP)	$= LF - ES - D = 0$

Floats

Total float (TT)	$= LF - ES - D$ or $LF - EF$
Free float (FF)	$= EF - ES - D$
Interfering float (INT.F)	$= TF - FF$
Independent float (IND.F)	$= EF - LS - D$

---

* Some computer application packages are programmed to start a project on day one, not zero, and to show the finish date at the end of an activity's last day, not the beginning of the next day.
Thus: $EF = ES + D - 1$ and $LS = LF - D + 1$.

selecting the longest path back through the network. Inspection of the network should now reveal the activities lying on the critical path. Critical activities can be identified as those having the same earliest and latest event times. The critical path is the longest path through the network, and in certain cases more than one critical path may be identified. The non-critical activities will have spare time or float, which can be determined for each of the non-critical activities. Figure 2.10 shows the different types of float associated with an activity. Critical activities have zero float, since any delay in these activities will affect the project duration. Critical path method definitions and equations are shown in Table 2.2.

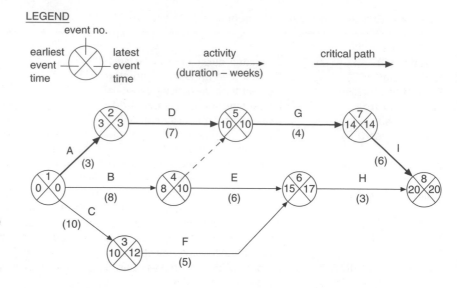

Fig. 2.11 An example of an analysed arrow diagram for a small project

Table 2.3 Calculated times for the network shown in Figure 2.11

Activity reference	Activity	Estimated duration (weeks)	Earliest start	Earliest finish	Latest start	Latest finish	Total float
1–2	A	3	0	3	0	3	0
1–4	B	8	0	8	2	10	2
1–3	C	10	0	10	2	12	2
2–5	D	7	3	10	3	10	0
4–5	dummy	0	8	8	10	10	2
4–6	E	6	8	14	11	17	3
3–6	F	5	10	15	12	17	2
5–7	G	4	10	14	10	14	0
6–8	H	3	15	18	17	20	2
7–8	I	6	14	20	14	20	0

Figure 2.11 illustrates an analysed network for a small project, with calculated start and finish times shown in Table 2.3.

Precedence diagrams

In addition to arrow diagrams, the precedence convention can be used to produce networks. This is an alternative critical path technique and is sometimes preferred to the arrow diagram format. In precedence diagrams, a node or box represents the activity, with arrows representing the logical relationships between activities. The size of the box bears no relationship to the duration of the activity, and the extent of information contained within the box is optional and dependent upon the level of detail required.

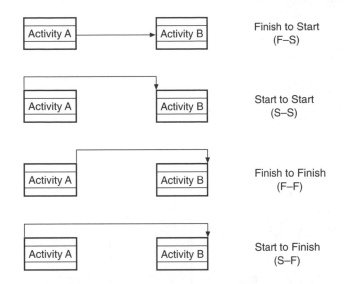

Fig. 2.12 Types of logic relationship between activities in precedence diagram format

The start of an activity is assumed to be the left side of the box, with the right side representing the finish. The relationships are therefore drawn in relation to starts and finishes of activities. These relationships can be categories as follows:

- *finish to start* (F–S): the succeeding activity cannot start until the preceding activity has finished;
- *start to start* (S–S): the start of the succeeding activity is dependent upon the start of the preceding activity;
- *finish to finish* (F–F): the finish of the succeeding activity is dependent upon the finish of the preceding activity;
- *start to finish* (S–F): the succeeding activity cannot finish until the preceding activity has started.

These relationships are shown in Figure 2.12 and are used appropriately to represent the logic of the network. Lead and lag times can be inserted on the link to indicate restraints on the relationships.

Precedence diagrams have several advantages over arrow diagrams. For example, dummy activities are not necessary to represent the logic and the overlapping of activities. Precedence diagrams are also easier to draw and modify, and additional activities can be inserted without changing node reference numbers.

Analysis of precedence diagrams is similar to that for arrow diagrams, and forward and backward passes are required to determine earliest and latest start and finish times. Float times are also calculated in the same way. Figure 2.13 shows an analysed network based on the precedence convention.

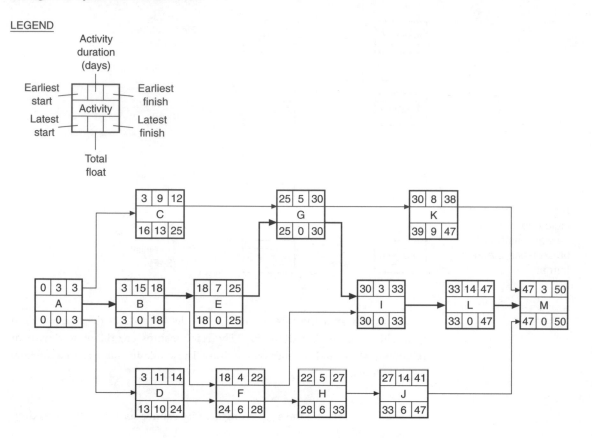

Fig. 2.13 An example of an analysed precedence diagram for a small project

Arrodence diagrams

The use of arrow diagrams or precedence diagrams to represent activities on a network is one of personal choice or may depend upon company policy. It could be argued, however, that the arrow diagram provides a more direct way of representing dependencies and restraints, whereas the precedence diagram, with nodes representing activities, is easier to construct and understand. There is less risk of making logical errors with precedence diagrams, since each activity is connected to others by a relationship. Additionally, all relevant information can be contained within the box.

Some disadvantages, however, do exist with precedence diagrams. On large networks in particular, drawing relationships becomes cumbersome, since relationships are not drawn across but between boxes. This can cause numerous lines to converge on certain parts of a diagram, making it difficult to follow interrelationships and dependencies (Lester, 1998).

Lester (1998) has identified problems with both arrow and precedence diagrams and as an alternative proposes the use of arrodence diagrams, which combine useful features from both techniques. In the arrodence diagram, the arrow diagram format is used and each activity is separated from the succeeding activity by a short arrow link. This format reduces the risk of making

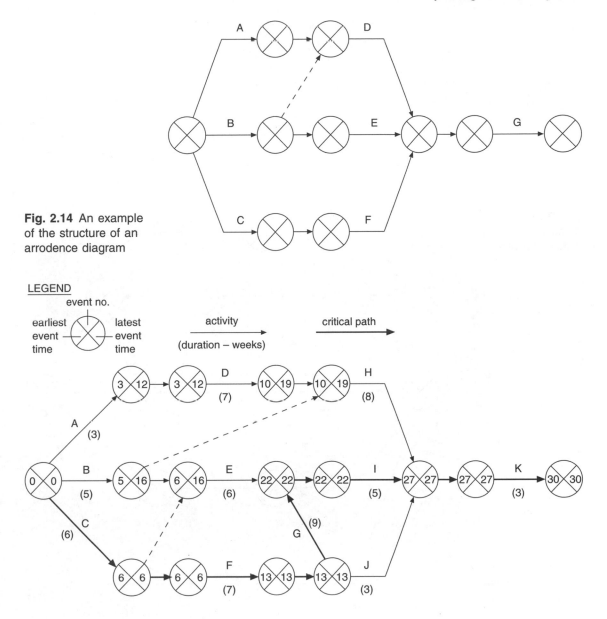

Fig. 2.14 An example of the structure of an arrodence diagram

Fig. 2.15 An example of an analysed arrodence diagram for a small project

logical errors while providing the clarity of links, and it allows float information to be as easily identified as with precedence diagrams. Figure 2.14 shows the structure of the arrodence format and the links between activities. Figure 2.15 shows an analysed arrodence diagram.

Complex networks   Complex projects give rise to comprehensive planning requirements. The time demands from clients, procurement issues, the extent of subcontract work and its co-ordination, together with other resource requirements during

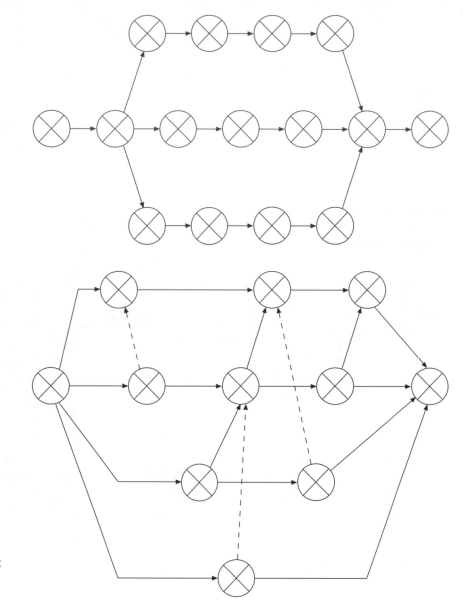

Fig. 2.16 A non-complex network of sixteen activities

Fig. 2.17 A complex network of sixteen activities

construction, add to the complexity. This complexity will also be reflected in the programming of activities.

The programming of a project which relates to similar work types may require a simple network and be relatively straightforward in terms of activity identification and logic. A new project, which may represent a new venture into certain types of work, will present more uncertainty and risk. Such a project will also require more thought and discussion in order to define activities and establish logic with dependencies and restraints (O'Brien, 1993). Figure 2.16 shows a non-complex network with few dependencies and restraints. Figure 2.17

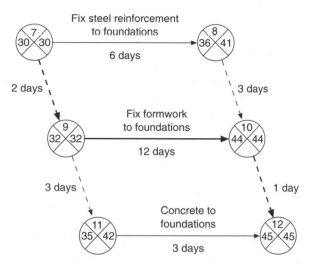

Fig. 2.18 An example of an analysed ladder construct

shows a more complex network with the same number of activities, but with more interconnections and requiring greater co-ordination of work activities.

Added complexity may also be introduced to a network regarding level of detail and the practical requirements of work activities. Typically, during construction, activities start shortly after preceding activities start. In arrow diagrams this relates to ladder constructs, where activities may be subdivided to allow the opportunity for this to be represented. In doing so, real-time dummies may be introduced to indicate lead and lag times (Oxley and Poskitt, 1996; Pilcher, 1992). Figure 2.18 shows a ladder construct incorporating real-time dummies. Several ladders may be nested into a single network. However, the level of detail and complexity of networks will be governed by the project itself and the requirements of work operations. This may also be a necessity to ensure logical sequences and the co-ordination of activities.

Sectionalised networks

When programming projects by network analysis, the size of the network will be influenced by the number of activities identified and included in the sequence logic and scheduling. On small projects, this may number up to 100 activities for the entire project and at this level may still be easy to read and interpret. On large projects, the work may be represented by several thousand activities, which would result in a complex diagram. To avoid such a situation, one technique is to produce a sectionalised network. This allows work to be segregated so that programmes can be easily read and understood. It also provides information in a way that shows how particular sections of work relate to other sections and the programme overall. On larger complex projects, several planners may have responsibility for specific sections of work on the same programme.

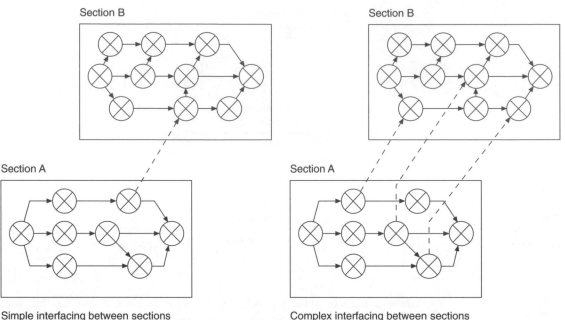

Simple interfacing between sections

Complex interfacing between sections

Fig. 2.19 Network sections with simple and complex interfacing

Sectionalisation can be achieved in several ways, through type of work, specialisation, specific trades, functionality, time-scale or stage of completion. Networks may also be sectionalised into cost centres or cost codes, which relate to a company's cost-control system. The usefulness of sectionalisation is that it provides the opportunity to focus on a particular set of activities. Ultimately, sectionalisation simplifies the network so that it can be used as an effective tool to assist in the production process and allow for effective monitoring and control of progress. Events in sectionalised arrow diagrams provide for the interfacing between sections, as shown in Figure 2.19.

Figure 2.20 shows the network for the construction of an industrial complex, where activities are grouped together to represent sections of the work related to the project. Interfacing is indicated where dependencies exist from other sections of work, the extent of which is dependent upon the activities and sequential logic of the network.

Sub-networks

On large projects, where different levels of detail may be required, sub-networks offer a means of representing work at a lower level of detail. They are extremely useful for enlarging a portion of a network for assessment purposes, and they allow a network to be kept down to controllable proportions. The application of sub-networks allows details of a project to be initially represented as an outline, where each section can be provided in greater detail at a lower level. The extent of levelling will be dependent upon the project and the extent of detail required in defining activities.

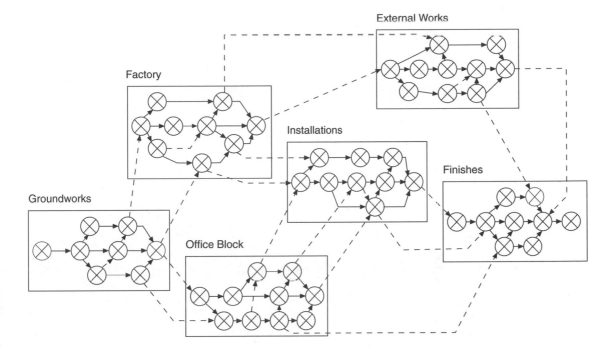

External Works

Factory

Installations

Finishes

Groundworks

Office Block

Fig. 2.20 An example of a sectionalised network for the construction of an industrial building complex

Figure 2.21 illustrates the sub-networks associated with part of a business centre development scheme, with sub-networks being produced at different levels of detail. On large complex projects with many activities, the top-level network may represent a complete construction project as one activity. The project itself may be represented by several hundred activities at various levels. The work within each sub-network is controlled within the network, possibly representing a particular part of a section of work.

As with sectionalised networks, interfacing is important to maintain logic and ensure that networks link together. Interfacing may be simple or complex. Simple interfacing involves the linking of networks at lower levels through one or two interfacing events. Complex interfacing, on the other hand, is where more than two interfaces are involved, and several interfaces may be included representing dependencies or restraints linking sub-networks and other sections of the network. While individual sub-networks can be analysed and have their own critical path, earliest and latest event dates have to be considered and may be a limiting factor on other networks. These limiting times on activities will have to be carried through via the interface in order that the affected sub-network calculations may be carried out. This situation will have a restraining effect on the networks concerned and will also influence the critical path. This may also have implications for resource requirements and subcontract work.

Clearly, on complex projects, there is a great need to ensure that the logic of the network and the dependencies are correct. Manual analysis and updating would prove difficult on such a network and is best carried out by computer.

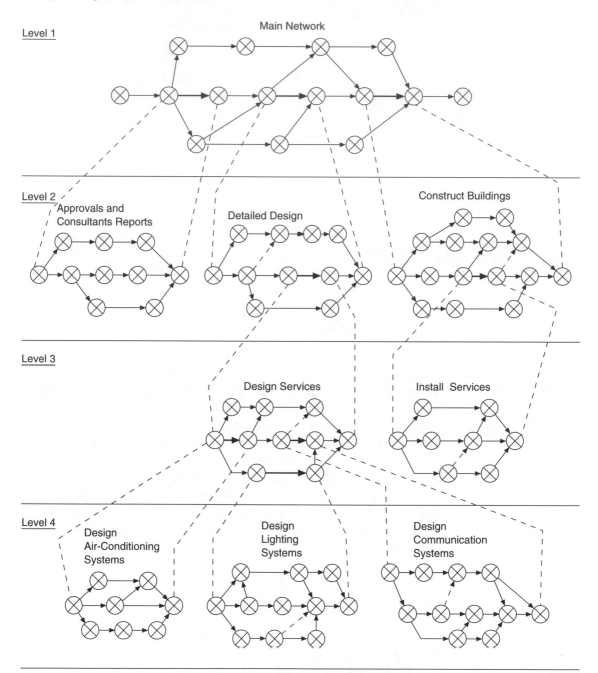

Fig. 2.21 An example of sub-networks for part of a business centre development scheme

In doing so, any changes that are made during the various stages of planning could be quickly analysed to ascertain the influence on other sub-networks and the likely impact on the completion of the project. Without the use of sub-networks on large complex projects, work would prove difficult to analyse and control.

Monitoring and controlling progress

The planning and programming of projects represent forward thinking in how work is to be carried out. This relates to the overall programme, in which several parties may have contributed and agreed in terms of phased completions and project duration. This process, however, becomes meaningless without the processes of monitoring and control. The established programme should be checked on a regular basis to look at what has been achieved and in order to establish future work requirements. This includes checking on physical progress in terms of work completed and resources expended; and identifying problem areas and slippages in progress in order that appropriate action can be taken.

The process of monitoring itself is concerned with an assessment of work completed over a particular time period. Work completed will have to be measured, and this may be carried out by the planner, who may be resident on site, or by site staff working on the project. It is essential that work completed to date is measured in a consistent manner which can be related to the programme. Information from an interim valuation or bonus payment scheme may not be appropriate, since ideally, progress should be related to the programme and the identified activities. Design forms are ideal, and companies may well have their own standard formats. Information can be transferred to the planning department, if carried out centrally, in order to assess progress. Several recording methods can be employed, including time measures of work completed, remaining duration or estimated percentage completion. Consistency is the important factor here, and this should be as accurate as possible to represent the current status of the project (Reiss, 1995).

Updated information will indicate progress since the last update and identify whether work is progressing to plan or whether slippages or delays have occurred. Additionally, other changes may occur through client requests, design changes, omissions or unforeseen items. New situations sometimes emerge, and control may be achieved through changes in logical sequences and establishing new critical paths. Additional constraints may require earlier completion dates and produce negative float on some activities.

From a practical perspective, the updating frequency should be timed at regular intervals and ideally coincide with site meetings, where affected parties can be briefed and remedial action taken as appropriate. The precise methods of reporting and controlling progress will depend to a large extent on the size of project and information requirements. Overall outline progress reports may be requested by senior management, whereas control of progress on site will have to relate to short-term planning stages, where detailed work activities can be identified together with related resources. It is therefore essential to establish effective systems of recording and reporting in order to effect the necessary control.

In monitoring progress, much will depend upon the method of programming used. Several progress-reporting methods can be adopted that are

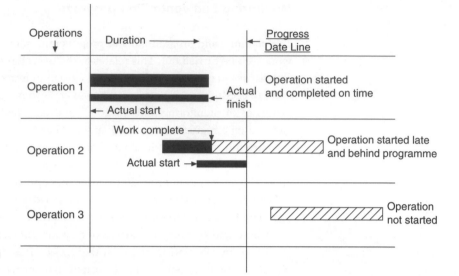

Fig. 2.22 An example of progressing work on a bar chart

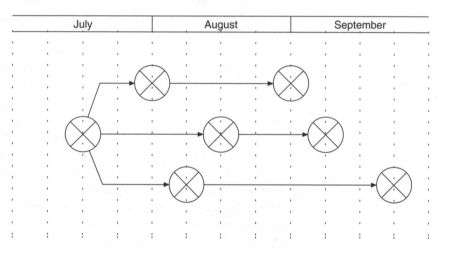

Fig. 2.23 Part of an arrow diagram drawn to a time-scale

commonly used in practice. If bar charts are being used, activities will be displayed horizontally together with float times. Additional bars can be displayed underneath the activity bar to represent the actual starts and completions of activities. Shading of the original bars can be carried out to indicate partial completion. Figure 2.22 illustrates this method using a bar chart.

Where arrow diagrams are employed, the network can be used to monitor progress directly. The logic diagram in its original form can be drawn to a time-scale, as shown in Figure 2.23 (O'Brien, 1993). Updated information is entered manually on the programme by shading, marking or colouring arrows and events to illustrate completed work and progress. Changes in durations of activities can be amended directly. Figure 2.24 shows these techniques for partly completed and amended activities. Parts of the network may have to be

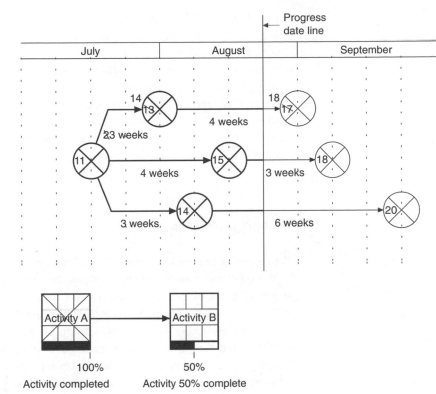

Fig. 2.24 Progressing work and updating information directly on a network

Fig. 2.25 Progressing activities on a precedence diagram

amended to cater for additional activities and changes in logical sequences. New critical paths may then be identified. Similar techniques can be employed with precedence diagrams and additional information entered in the nodes for each activity, as shown in Figure 2.25 (Lester, 1991).

A simple and effective system of monitoring and recording progress will prove useful for site management and operatives. Slippages can be identified together with possible future problem areas, allowing control procedures to be introduced as appropriate.

The programme used for monitoring and controlling progress can also represent a useful document when cases arise involving the preparation of claims. The contractor will have a stronger case, particularly if the programme has been agreed along with the methods of construction. This can act as supporting evidence in a dispute situation. The contractor in carrying out the works should be supplied with all information required, including design details and other instructions, and information on practical items such as possession of certain parts of the site, access and equipment, which may have been stated in the tender documents, in addition to information related to nominated subcontractors. The contractor may also have constraints placed on the project by the client requesting early completion or occupation of parts of the project (Lester, 1991).

Effective monitoring procedures in these situations are crucial in order to highlight delays and slippages. Failure to provide information or changes imposed on a contractor may result in disruption to the sequence of activities. This in turn may have a knock-on effect on other work, causing further delays, problems of resource management and increased cost. This may also influence the overall project duration and warrant an extension of time. Monitoring the work and recording actual start and finish times, including the effect on float, should indicate the activities affected and the impact on time and cost. As-built networks will prove advantageous, if time for their preparation permits, and such recorded information will add strength to the case for a contractor and be supportive in any dispute situation (Reiss, 1995).

Managing resources

The resources required to carry out activities on a project will be determined by the type and quantity of work. Resources represent the workforce on a project in terms of the various trades and operatives. Plant and equipment, including materials, components and finance requirements, are also representative of resources. Various combination of resources can be assigned to activities, and this is normally considered during the tender period when determining the duration of activities or when preparing a method statement.

Resource allocation may be further considered by production staff during construction. Resources selected for estimating purposes may be changed and selected by production staff based on previous experience. This is perfectly acceptable providing that estimated allowances are not exceeded. At the outset, such selection will have to be decided for each activity, since no account may have been taken of availability. However, resource aggregation, where the total number of resources available is considered, will allow the estimated resources to be compared with those allocated.

The process of resource levelling will minimise fluctuations in resource usage and ensure that the resources available are not exceeded. If this proves problematic, activities will need to be delayed until resources are available. The impact here on critical activities is important since this will influence project duration.

Project duration may also be affected by the process of resource smoothing. High and low demand for resources can be smoothed out in an attempt to provide resource continuity and utilisation. Situations of having labour and plant repeatedly on and off site should be avoided if at all possible. Making adjustments to the planning of starting and finishing times of non-critical activities within their float limitations may provide solutions and avoid excessive resource demand (Oxley and Poskitt, 1996). This is illustrated in Figures 2.26 and 2.27. However, ideal solutions may be difficult to find considering the practicalities of construction and the demand for access, working space and the co-ordination of trades. Additional considerations may therefore be necessary by changing the logical sequences of activities or alternative

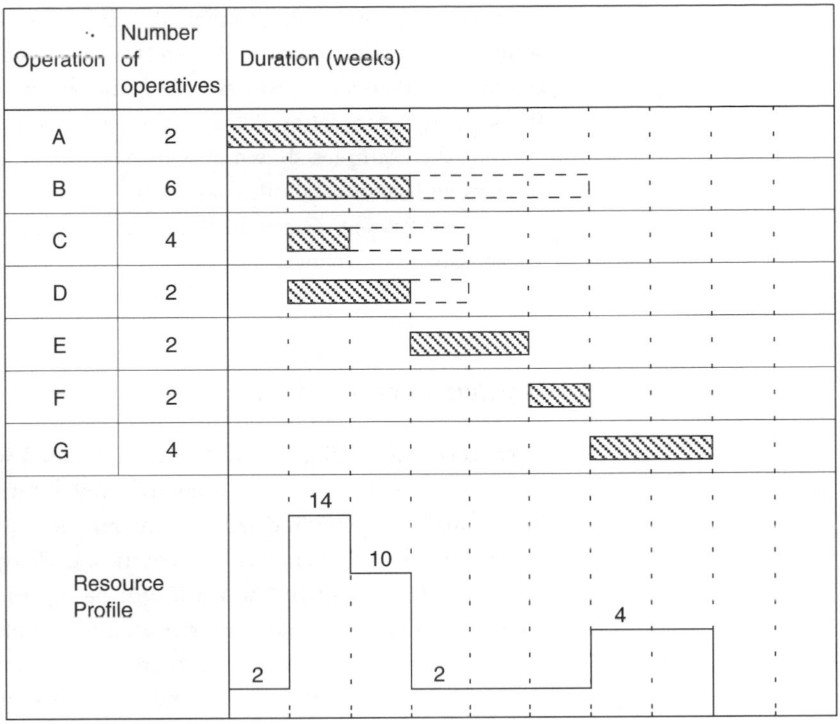

Fig. 2.26 An example of a resource profile for activities

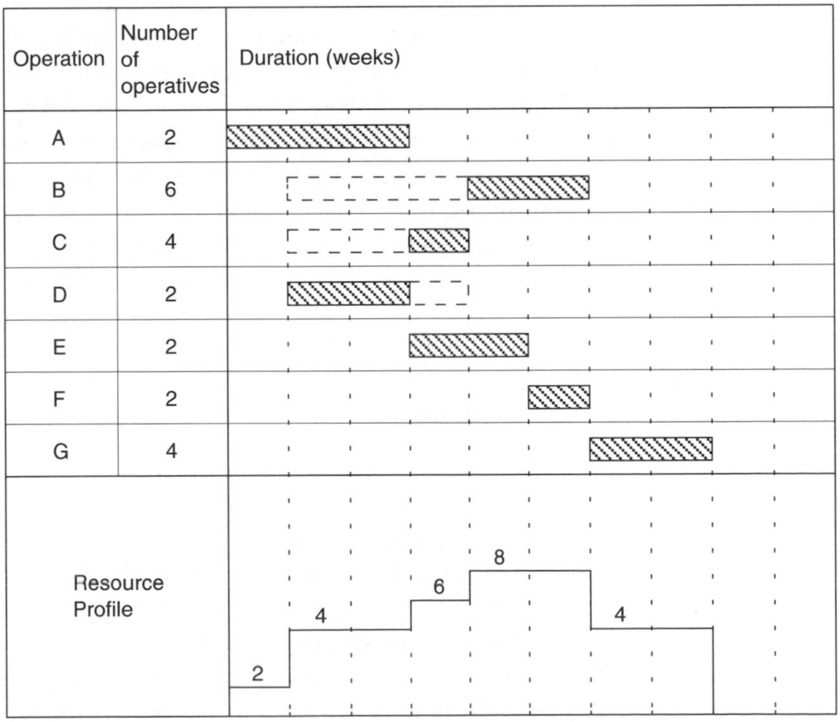

Fig. 2.27 An example of resource smoothing by using float for non-critical activities

resource allocation. Trade-offs may also have to be considered and an optimum solution eventually decided. The effective management and high utilisation of resources is therefore a concern for management in order to ensure activity times, project completion dates and budgets are not exceeded.

While the examples shown in Figures 2.26 and 2.27 are simplistic, on large projects with many activities, solutions are best arrived at by computer. Different scenarios can be tested before decisions are made. This will also include an assessment of financial resources for the consideration of cash flow.

Time/cost optimisation

Time/cost optimisation is a technique associated with network analysis. It is used in order to find the most economic way of carrying out a project in terms of time and cost. The time aspect is the representation of a prepared network indicating a logical sequence of activities, each with its own duration. The longest path through the network will be represented by the critical path, which determines the total time required to complete the project.

The cost of a project can be considered as a combination of direct and indirect costs. The direct costs relate to the resources required to complete activities in terms of labour, plant, equipment, materials and components. Indirect costs relate to items such as on-site costs and operatives' wages, many of which are representative of time-related costs. Additionally, contractual costs may be incurred through penalties for late completion and are also calculated on a time-related basis. The variation in duration of critical activities will therefore affect the cost of these activities and the total cost of a project (Pilcher, 1992).

To accelerate activities is more expensive in terms of direct costs. As activity duration decreases, costs increase. With indirect costs, as project duration increases so will total indirect costs. Time/cost optimisation is therefore concerned with investigating various scenarios and establishing project duration at minimum cost. There may be several reasons for wanting to achieving this:

- to complete work by a programmed date;
- to complete work on an earlier date requested by the client;
- to recover from slippage in a programme.

Reducing the duration of an activity can be achieved in several ways, including increasing resources allocated to an activity, or possibly working overtime or working additional shifts. The reduction in duration is termed 'crashing' an activity to establish the 'crash time', for which there will also be an associated 'crash cost'.

One may consider, therefore, that to achieve minimum duration crashing all activities would be the most effective solution. This can be achieved by the addition of resources for activities or by putting the entire site on overtime.

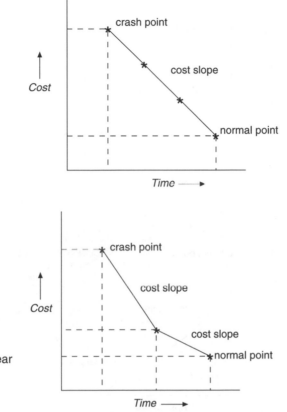

Fig. 2.28 A linear relationship between time and cost for an activity

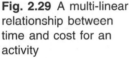

Fig. 2.29 A multi-linear relationship between time and cost for an activity

However, in practice this may not be the ideal and most cost-effective solution. If a decision is made to do this, it may be at a time when labour and plant resources are at a high level. Work carried out on non-critical activities will not shorten the project duration but will increase costs. To avoid this, selectivity needs to be applied and individual activities investigated in relation to time and cost (O'Brien, 1993).

When considering the relationship between time and cost, various relationships exist (Ahuja *et al.*, 1994). Figure 2.28 shows a linear relationship associated with activities. For a reduction in duration a linear increase in cost occurs. A cost slope for activities can therefore be determined as follows:

cost slope = (crash cost − normal cost) / (normal duration − crash duration)

A particular duration for an activity along the cost slope can therefore be selected with a corresponding cost.

Additionally, a multi-linear relationship may exist where an activity has a linear relationship up to a certain point and thereafter the cost slope changes. This may be due to specific resources being required to reduce duration. This multi-linear relationship is shown in Figure 2.29.

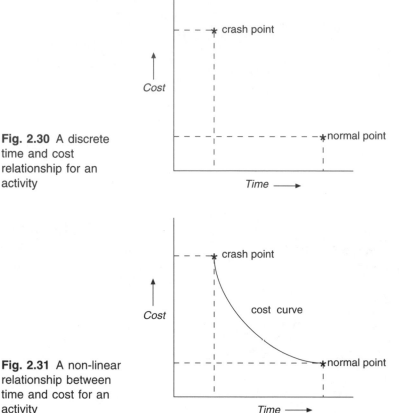

Fig. 2.30 A discrete time and cost relationship for an activity

Fig. 2.31 A non-linear relationship between time and cost for an activity

Situations may also arise where discrete costs occur. It may be that specific plant or equipment is required to reduce duration, incurring a lump sum cost. In this case there is no cost slope, since no intermediate point can be selected. If such an activity is selected for crashing, the situation is the normal cost or the crash cost. This situation is illustrated in Figure 2.30.

A non-linear relationship may also exist between normal and crash situations on certain activities. This is shown in Figure 2.31 and may occur where different combinations of resources produces a non-linear relationship. Combining direct and indirect cost curves for crashed activities would establish total costs, from which the low point on the total cost curve would determine the least cost for optimum project duration. This is shown in Figure 2.32.

All these different relationships may occur on the same project, and each situation would have to be considered individually. In practice, the reduction of duration would have to be applied carefully. Increasing resources to reduce duration can only be done up to a certain limit from the practicalities of construction. At some stage, a point of diminishing returns would be reached where resources could not work effectively, so selecting the crash duration has to be done realistically.

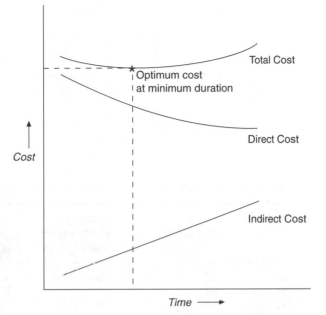

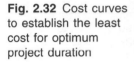

Fig. 2.32 Cost curves to establish the least cost for optimum project duration

The process of time/cost optimisation is dependent upon the project network and cost information available. Several situations may arise involving both compression and complex compression (Oxley and Poskitt, 1996). Compression may take place involving several activities with cost slopes identified on a network. Re-analysis would be carried out at each stage to arrive at the appropriate duration. In some situations, however, it may be possible to carry out a number of compressions involving several activities with linear, multi-linear or discrete cost curves. This would result in a more detailed analysis and represent complex compression. Consideration would also have to be given to situations where more than one critical path emerges. In such cases, further compressions would have to be carried out on these paths concurrently.

While compression of activities is aimed at the reduction of duration for a project, certain steps can be taken to reduce costs further. This can be achieved by decompressing non-critical activities, thereby increasing their duration and decreasing cost. This may be achieved by reducing the resources on the activity and allowing the activity to utilise its float. Resources allocated for the normal times may be transferred to other activities or activities on the critical path. Additionally, activities which have not been started may have their resources built up progressively. Decompression of an activity can be carried out until the activity becomes critical. However, the extent of decompression applied to activities would be dependent upon the practicalities of reduced resources being able to carry out the work.

In carrying out time/cost optimisation, a selective process takes place to decide which activities to compress or decompress. On small networks this

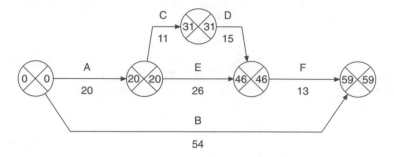

Fig. 2.33 Part of an analysed network for compression

Table 2.4 Cost information associated with the network shown in Figure 2.33

Activity	Duration (days)	Maximum compression (days)	Normal cost (£)	Cost slope (£)
A	20	5	4,700	20
B	54	–	5,485	–
C	11	4	6,536	first 3 days £35 per day, £85 per day thereafter
D	15	5	3,400	first 2 days £40 per day, £95 per day thereafter
E	26	10	3,950	55
F	Method (1) 20	10	1,100	90
	Method (2) 13	3	1,400	150

may be relatively straightforward, but on large complex networks this can be more problematic. In such cases, a more analytical approach to activity selection involving a critical theorem can be employed (Ahuja *et al.*, 1994). In order to carry out time/cost optimisation, certain cost information will need to be collected. In practice, this may prove difficult if cost systems do not relate to construction activities (O'Brien, 1993). However, if this is not the case, information can be tabulated for normal and crash durations and costs. Critical activities can be compressed in the order of least cost slopes. Non-critical activities can be investigated for possible decompression in the order of greatest cost slope. Direct and indirect costs can then be calculated for each project duration. Adding these two costs together will enable the optimum cost of the project to be determined.

To illustrate the process of compression, part of a network shown in Figure 2.33 has been compressed using the information in Table 2.4.

first compression
Compress activity A by 5 days
$$5 \times £20 = £100$$

second compression
Compress activity C by 3 days
$$3 \times £35 = £105$$
Compress activity E by 3 days
$$3 \times £55 = \underline{£165}$$
$$\text{Total} \quad £270$$

third compression
Compress activity D by 2 days
$$2 \times £40 = £80$$
Compress activity E by 2 days
$$2 \times £55 = \underline{£110}$$
$$\text{Total} \quad £190$$

fourth compression
Compress activity E by 1 day
$$1 \times £55 = £55$$
Compress activity C by 1 day
$$1 \times £85 = \underline{£85}$$
$$\text{Total} \quad £140$$

fifth compression
Compress activity E by 3 days
$$3 \times £55 = £165$$
Compress activity D by 3 days
$$3 \times £95 = \underline{£285}$$
$$\text{Total} \quad £450$$

sixth compression
Method (2) selected
Compress activity F by 3 days
$$3 \times £150 = £450$$

To optimise costs, direct costs and indirect costs must be combined. The direct costs are £120 per day. Details of compressions including costs and durations are summarised in Table 2.5. The optimum cost is £31,911, with a duration of 49 days. Costs may be reduced further by decompressing activity B.

Multi-project planning

The planning of individual projects can be carried out by the application of programming techniques described earlier. In practice, however, situations often arise where several projects have to be constructed simultaneously. This applies equally to both large developments involving several contractors and

Table 2.5 Summary of time and cost information for compressing the network shown in Figure 2.33

Compression	Compression cost (£)	Normal cost (£)	Direct cost (£)	Indirect cost (£)	Total cost (£)	Duration (days)
Normal	–	25,471	25,471	7,080	32,551	59
First compression	100		25,571	6,480	32,051	54
Second compression	270		25,841	6,120	31,961	51
Third compression	190		26,031	5,880	31,911	49
Fourth compression	140		26,171	5,760	31,931	48
Fifth compression	450		26,621	5,400	32,021	45
Sixth compression	450		27,071	5,040	32,111	42

small projects constructed by a single contractor. Situations may occur involving developments where several identifiable projects relate to some overall development scheme. Several contractors may have been awarded contracts, and a project team may have the responsibility of co-ordinating the entire development to meet a specified completion date.

The network techniques described earlier can be utilised in such cases, resulting in a large complex network. Figure 2.34 shows the network for a city development scheme where several projects need to be constructed simultaneously. These individual networks, or fragnets (fragmented networks), show project activities which are linked together through interfaces indicating certain restraints and dependencies. The network for each project will have its own critical path and resource requirements, but additionally a critical path for the entire development can be determined.

Monitoring and controlling such a development will obviously be a major task, and progressing the projects and providing updates would be best dealt with by computer. Plans would have to be merged and some effective system of monitoring established. Many items would therefore need consideration on such a development in order to bring it to a successful completion.

On a smaller scale, the situation also exists for the small contractor where several projects may been awarded with staggered start and completion times. This may be in addition to a contractor's on-going projects. The problem here, for the contractor, is that the projects may also have to be serviced from a common pool of resources in terms of labour and plant. The programming and management of resources therefore becomes a major problem where work continuity and high utilisation of resources become real issues (Reiss, 1996).

Figure 2.35 shows bar charts for several small projects being constructed simultaneously and where activities have been planned in order to provide for work continuity. At the planning stage, consideration should be given to the continuity of work for labour and plant from one project to another. Where possible it is preferable to avoid leaving a project unmanned, and subcontract

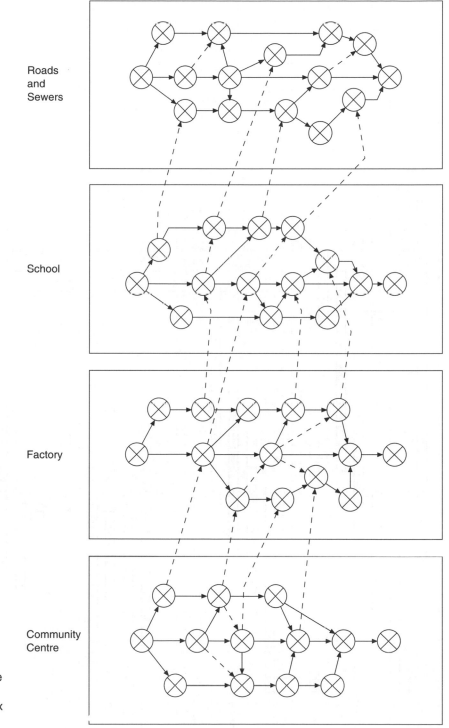

Fig. 2.34 An example of multi-project planning with complex interfacing

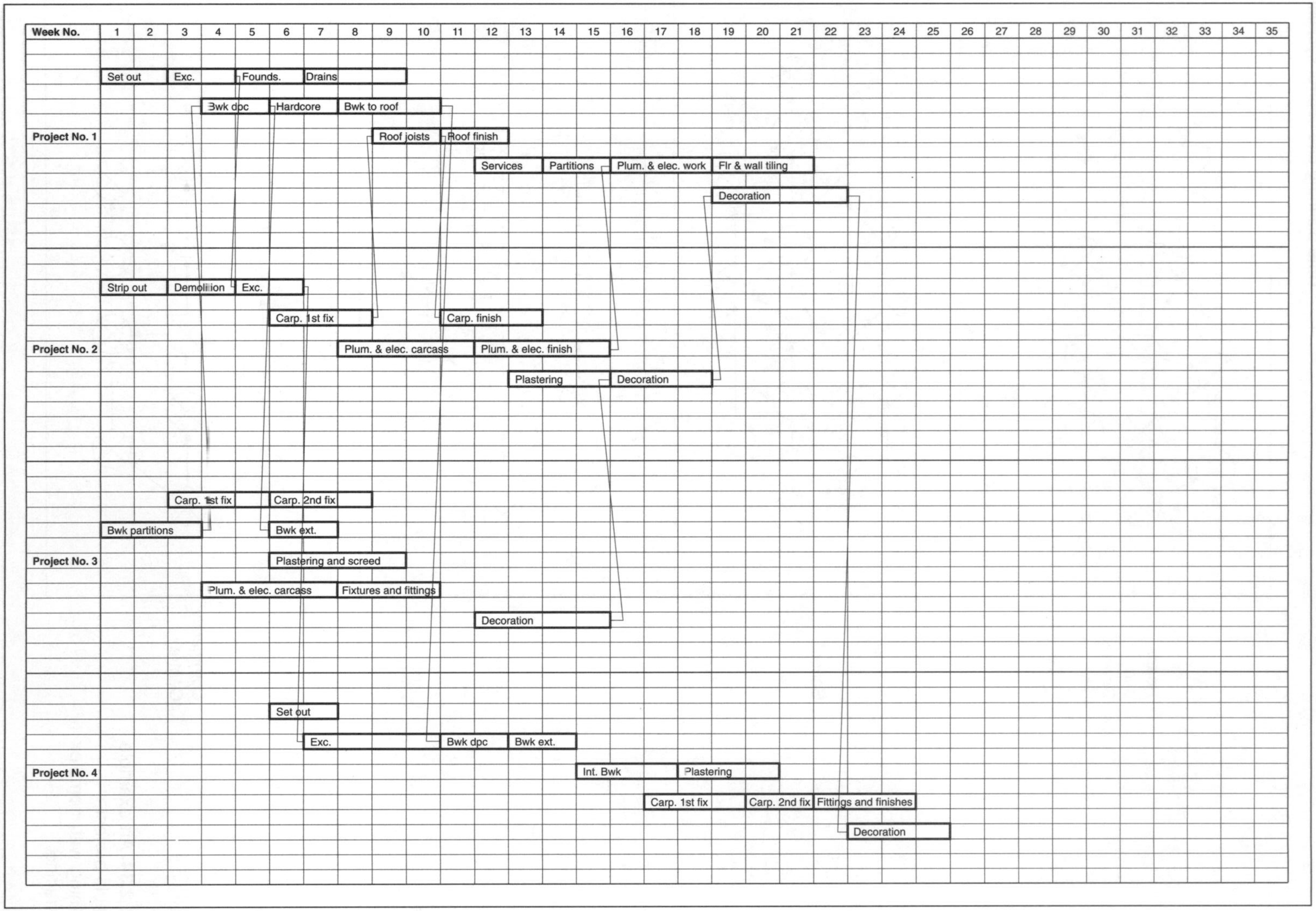

Fig. 2.35 Multi-project planning using bar charts to illustrate links for work continuity

labour should be integrated into the programme of work as effectively as possible. Material resources are also a major concern, and deliveries should be scheduled to coincide with operatives being on site. During construction, the monitoring of progress will be important, particularly with reference to effects on other projects. Subcontractors should be kept fully informed, and access and working space should be made available to avoid conflict of trades. Plant and equipment should also be monitored and be available from previous sites as required.

There are no instant solutions to these situations in multi-project planning. Such situations in practice are rarely problem-free, but utilising appropriate planning and programming techniques will at least alert management to potential problem areas.

Computer software for planning and control

The availability of software for planning and control is now quite extensive, and many application packages are available to run on various hardware platforms (CICA; CIOB). Advances in software development have resulted in application packages being user-friendly, with the provision of a graphical interface between the user and the computer (Paulson, 1995).

Several packages allow bar charts to be drawn directly on the screen, together with the insertion of links between activities. Other application packages allow networks to be drawn directly on the screen in precedence or arrow formats, with some packages supporting both conventions. The cheapness and availability of powerful personal computers which run most packages mean that users can be selective in the software they wish to use.

Software considerations Before purchasing an application package, it is recommended that users first establish criteria for selection. These should include all the features required from a package in addition to user documentation, training and the support available from vendors. The facilities offered by planning software are expanding rapidly, and further developments are being made continuously (Tulip, 1998). It is therefore important to consider new features and their usefulness. The sophistication of features will often be reflected in the price of a package, so users should be clear at the outset what their requirements are before purchase.

Some common facilities offer by planning software include:

- bar chart and network displays
- project calendars
- resource calendars

- organisation of activities
- resource profiles and tables
- work breakdown structures
- creating and merging sub-projects
- monitoring and updating
- report generation
- customisation of project information
- remote project data communications through e-mail systems
- links to other software systems.

The great advantage of planning software is the speed of data processing and information generation. Once project data have been entered into a system this can then be interrogated and different scenarios investigated to determine likely outcomes on which to base decisions.

One important point here is the thought process which is required to establish network logic and the sequence of activities. This is the task of the planner, not the computer. While knowledge-based systems and advances in software are taking place all the time, the preparation of a network is a manual process. The computer is only a tool to provide rapid processing and project information. This is particularly important where several parties are involved in a project and where collective decisions can be made prior to agreements being reached. Once agreed, data can be input into the computer and the facilities of the package utilised.

Project-planning example using Primavera SureTrak

To illustrate some of the features of planning software, Primavera SureTrak has been used to plan a small project, as shown in Figure 2.36. Additional details are given in Table 2.6. SureTrak provides a useful interactive graphical environment in which to plan, schedule and control projects. The package can be used for overall to short-term planning. This also includes projects of any size from small works to large-scale developments comprising several thousands of activities and resources (SureTrak Project Manager 2.0, 1997).

A project can be set up initially using the data input screen shown in Figure 2.37. The name of the project and the title can be entered in addition to other information, including project start and finish times. Calendars can be established initially from a global calendar to reflect different work periods, as shown in Figure 2.38. These would include normal Monday-to-Friday working with specific working hours in a day; seven-day, 24-hour working; or weekend working only. Calendars are established to suit a project and particular activities and resources.

Activities can be entered in the columns of the bar chart display or via the activity form, as shown in Figure 2.39. Relationships can be established directly on the screen using a graphical tool or details entered in the successor or predecessor dialog box. Relationships can be displayed by linking arrows

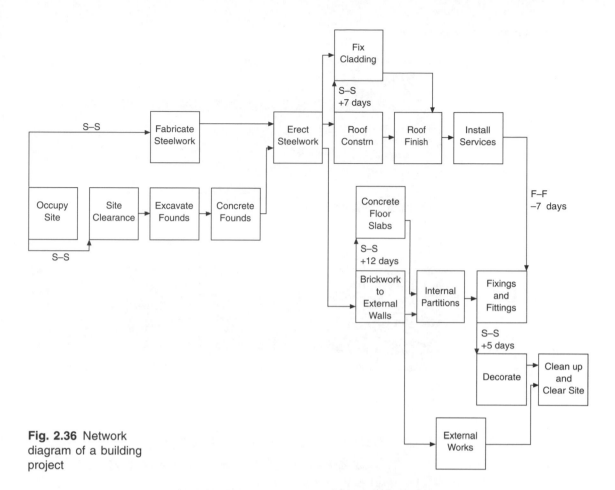

Fig. 2.36 Network diagram of a building project

Table 2.6 Estimated durations and resource allocation details for building project activities

Activity description	Estimated duration (days)	Resources allocated
Occupy site	0	£120,000
Site clearance	14	4 labourers
Excavate foundations	12	4 labourers
Concrete foundations	7	4 labourers
Fabricate steelwork	25	2 steel erectors
Erect steelwork	14	4 steel erectors
Fix cladding	12	3 labourers
Roof construction	10	3 carpenters
Roof finish	7	3 labourers
Install services	46	4 service engineers
Concrete floor slabs	15	4 labourers
Brickwork to external walls	35	4 bricklayers
Internal partitions	14	2 bricklayers
Fixings and fittings	21	2 carpenters
Decorate	21	4 painters
External works	28	4 labourers
Clean up and clear site	10	4 labourers

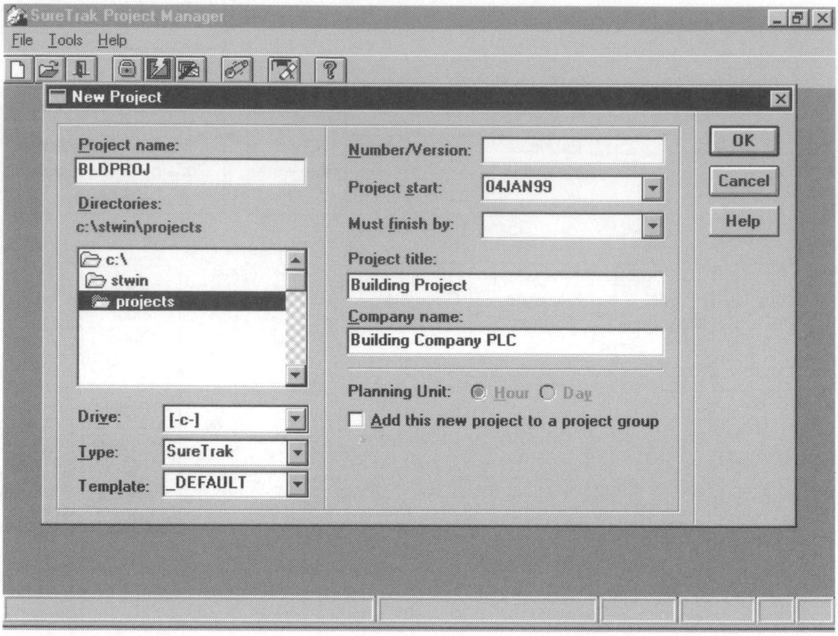

Fig. 2.37 Entering project information for a new project

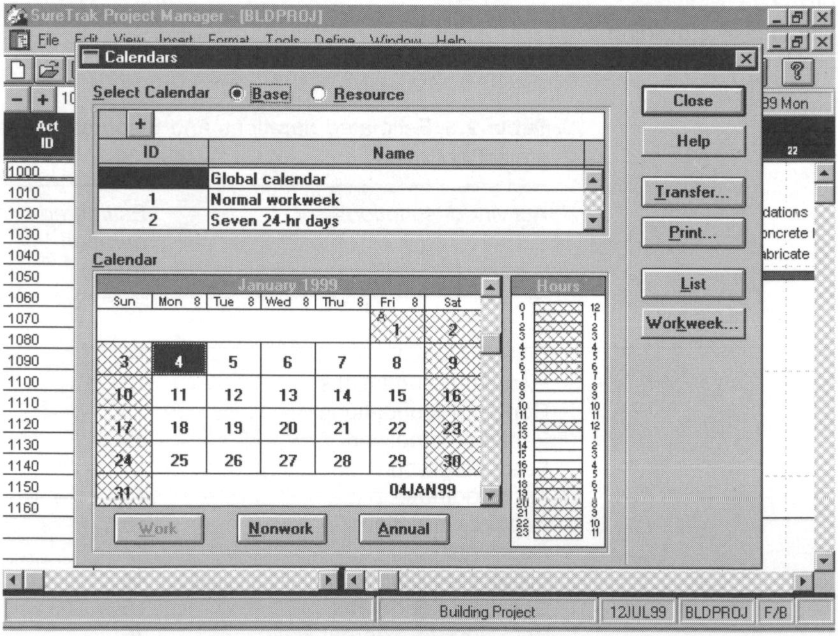

Fig. 2.38 Defining calendars for a project

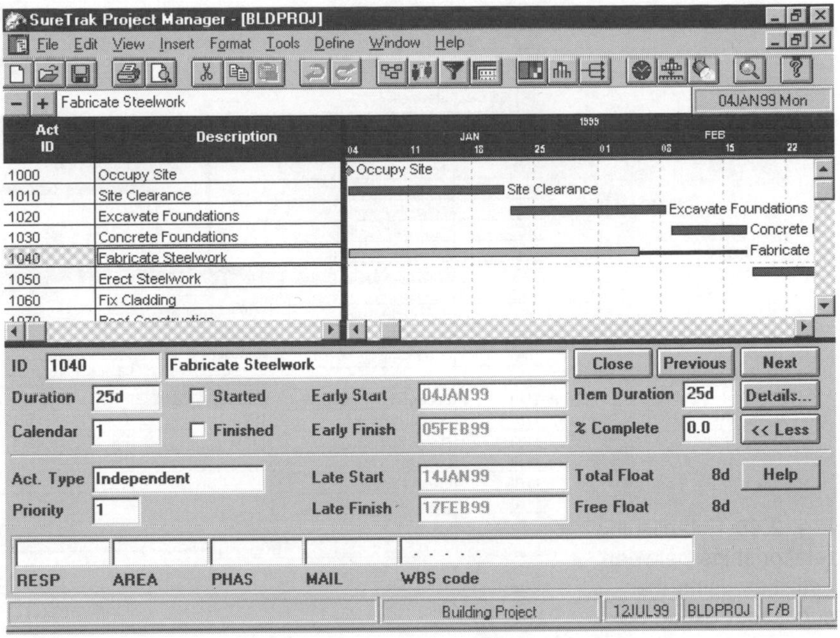

Fig. 2.39 Entering activity details on the activity form

between activities, as shown in Figure 2.40. Constraints can also be applied to individual activities in a variety of ways. Figure 2.41 shows the constraints dialog box, where types of constraint can be entered as required. Constraints imposed on a project may have significant influence, and situations may result where activities indicate negative float.

Resources used on a project in connection with activities have first to be defined. Details of cost and revenue generated can also be entered, together with resource limits and availability. As availability of a resource may change over the duration of a project, different dates can be catered for by making individual entries. The define resources dialog box is shown in Figure 2.42. Once defined, resources can be allocated to activities, as shown in Figure 2.43. The resource and the number of units can be entered in the resources dialog box and several resources entered against each activity.

The extent of resource allocation can be examined by use of the resource profile facility. This shows individual resources or selected resources and the amount of resource related to activities and time periods, together with cumulative quantities. Particular resources exceeding their set limit can be identified, as shown in Figure 2.44. This provides the opportunity to increase the limit of the resource or decide on some other course of action, which may include resource smoothing within the float available on non-critical activities. In addition to the resource profile display, resources can also be viewed by means of resource tables. Selected resources or totals can be chosen, and cost can be viewed as opposed to quantity, as shown in Figure 2.45.

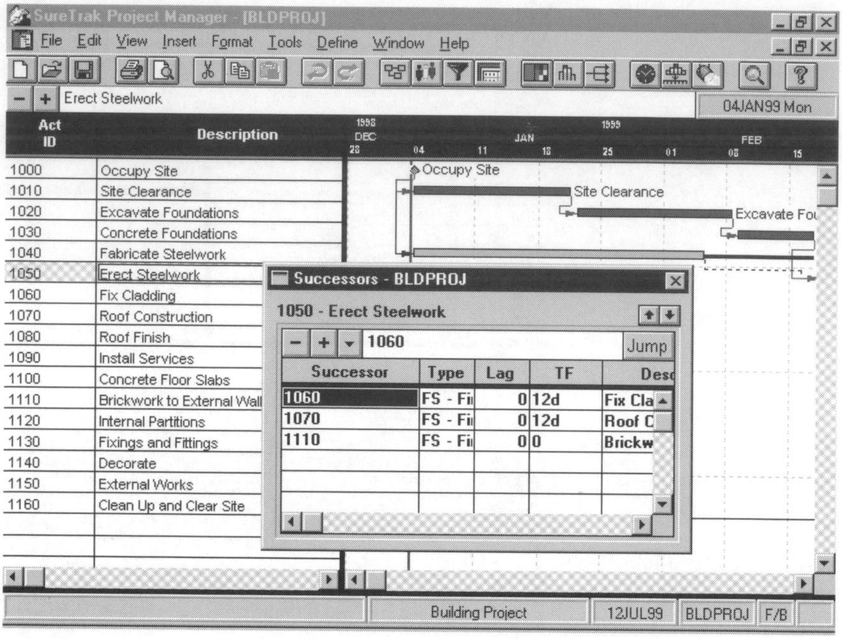

Fig. 2.40 Establishing relationships between activities using the successor dialog box

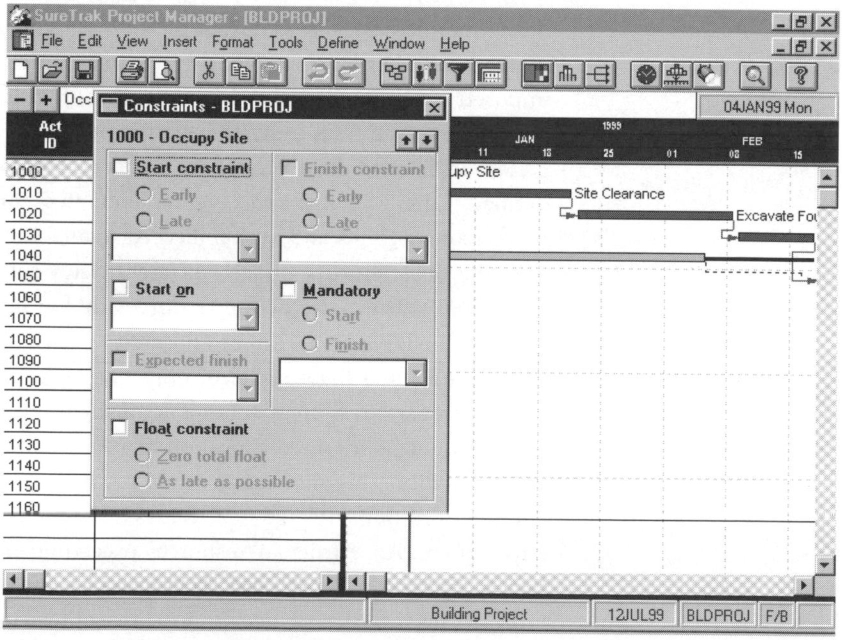

Fig. 2.41 Dialog box for entering constraints on an activity

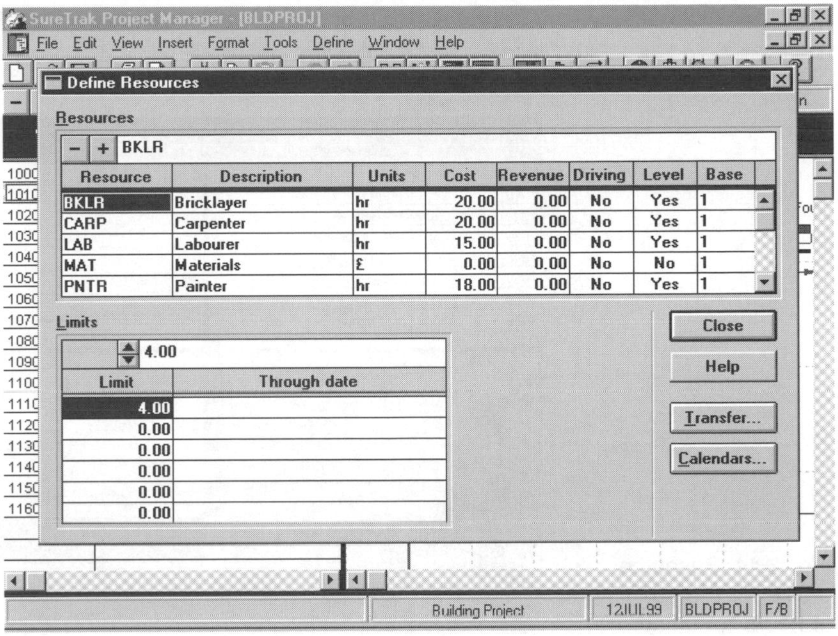

Fig. 2.42 Defining resources for a project

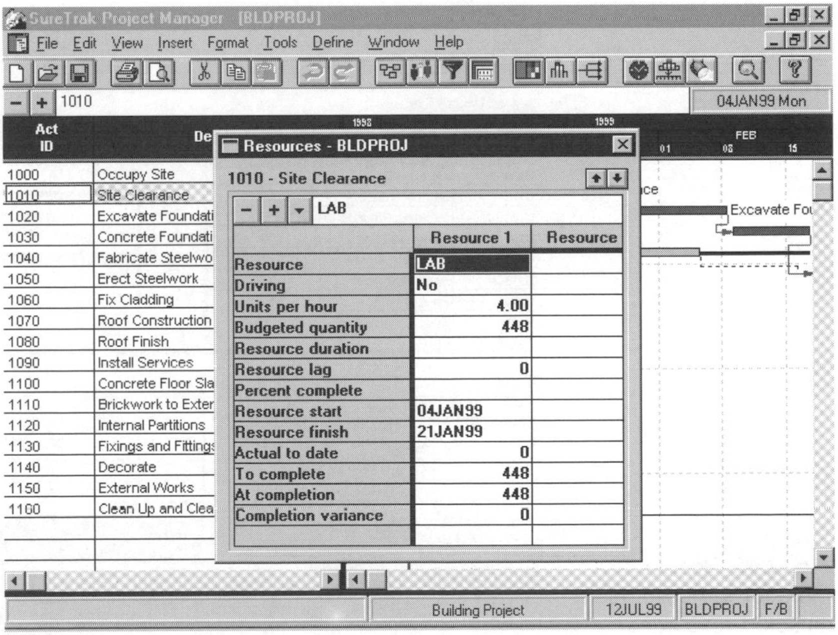

Fig. 2.43 Allocating resources to activities

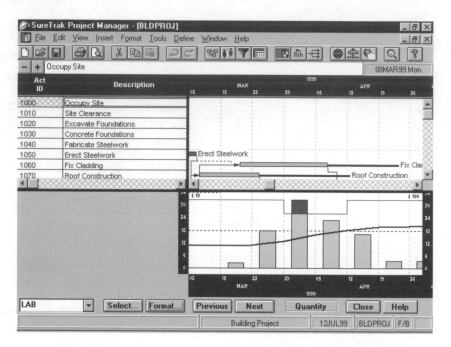

Fig. 2.44 Resource profile for the quantity of hours for labourers

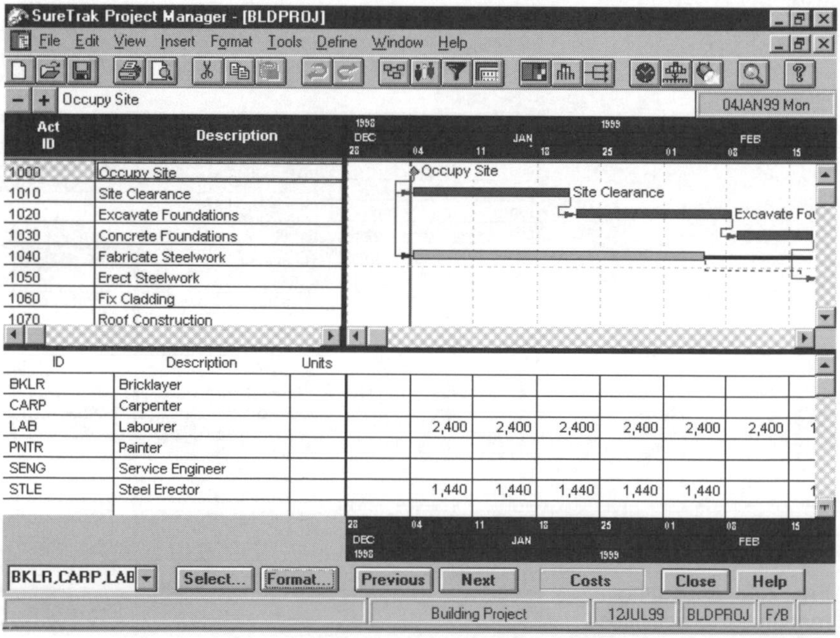

Fig. 2.45 Resource table for the costs of selected resources

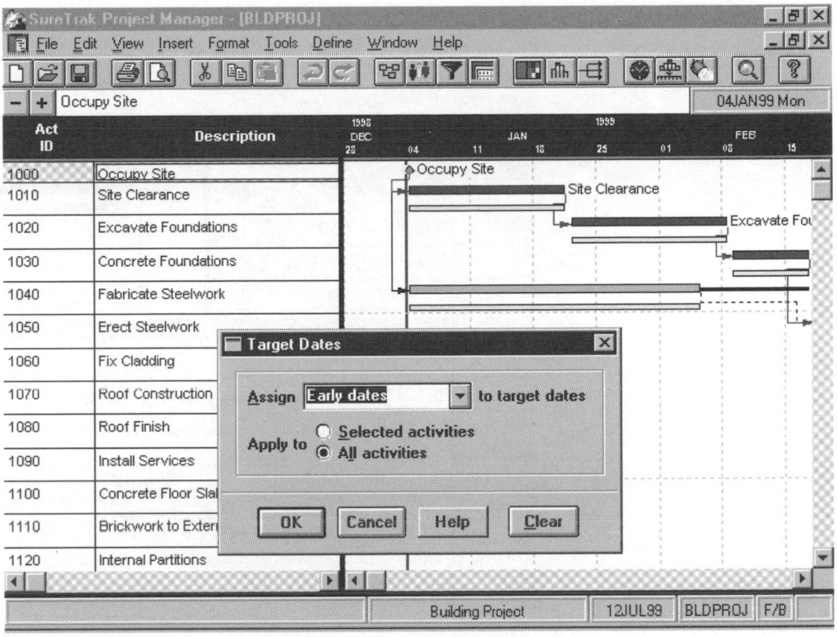

Fig. 2.46 Assigning target dates to activities

For monitoring progress on projects, target dates can be established. Dates can be assigned to activities via the target dates dialog box, and target bars made visible under the main bars, as shown in Figure 2.46. To update progress, the data date line can be dragged or set as appropriate for the period of progress. This process highlights the bar chart, together with highlighting the rows of relevant activities. Details can be entered for each activity in turn with regard to progress made. When update is selected, the main bar is updated to reflect actual start and finish times. This allows inspection of the bar chart to view changes that may have occurred from the target bars, as shown in Figure 2.47.

In addition to the bar chart display, network representation can also be used with SureTrak. The activity box can be configured to show particular types of detail required, as shown in Figure 2.48, and the network can be drawn on the screen and relationships entered as appropriate. A network can therefore be built up progressively, and while only part of the network may be visible on the screen, a cosmic view can be provided with a highlighted view area of specific activities. This is illustrated in Figure 2.49.

Progress of activities is also indicated in the PERT view, and completed activities are shown with a cross for completion and a single diagonal line for work having started on an activity, as shown in Figure 2.50.

While only a glimpse of SureTrak's facilities has been illustrated, many other features are available within the package. SureTrak is a powerful planning tool, with comprehensive screen displays, information representation and report-generating capabilities. The package is highly popular and is used extensively in the construction industry. This example has provided an indication of some of the features available in project-planning software to plan and control projects.

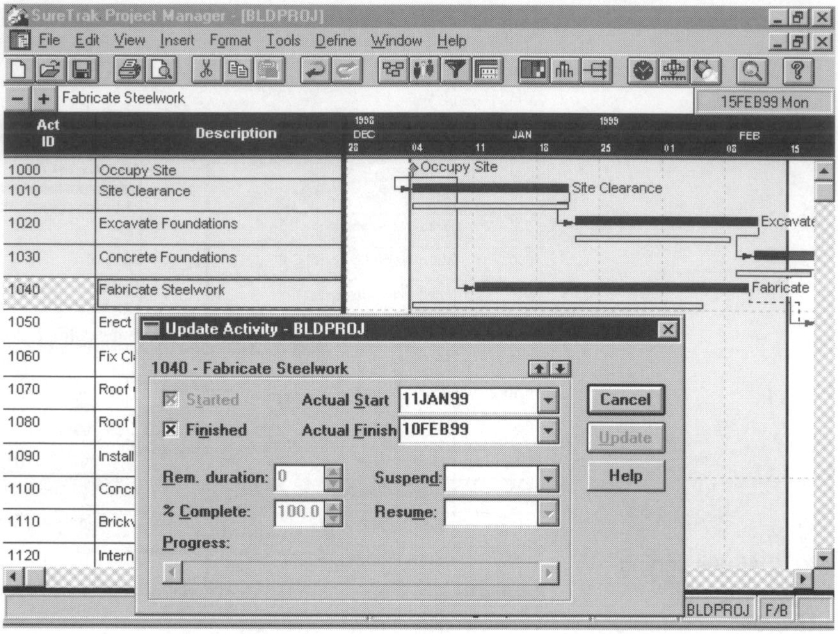

Fig. 2.47 Updating details of an activity

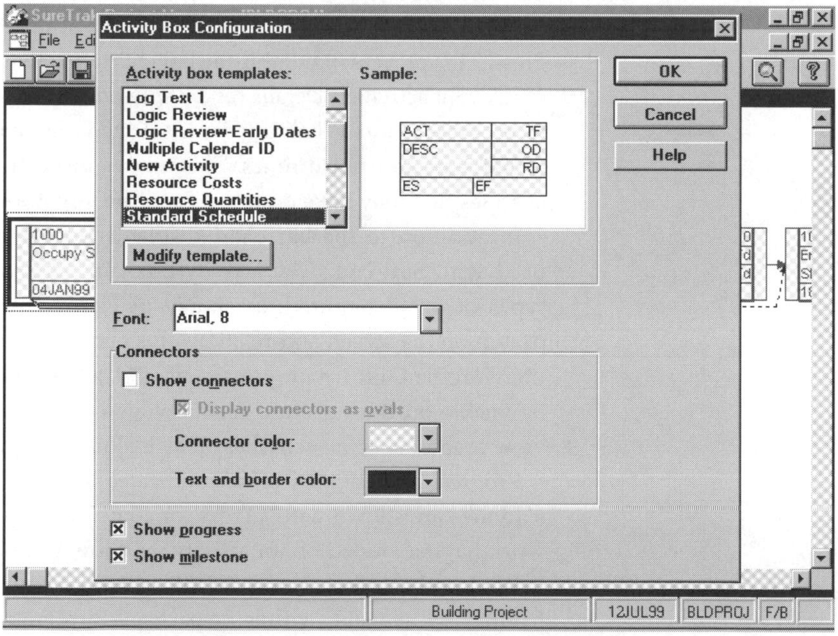

Fig. 2.48 Selecting activity box configuration

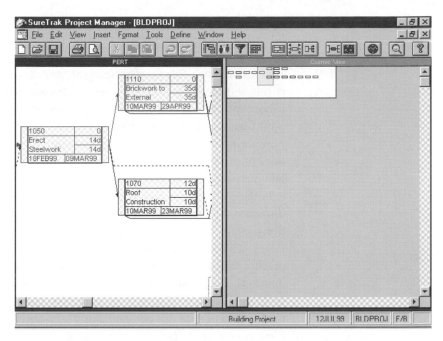

Fig. 2.49 Selected display of activities with a cosmic view of a network

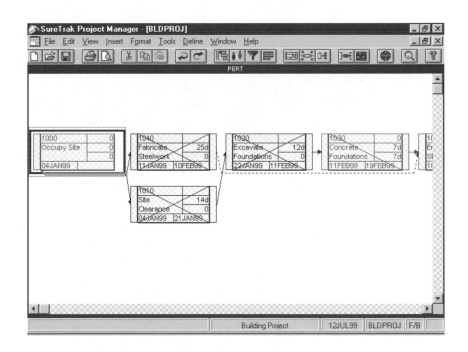

Fig. 2.50 Selected section of network showing progress on activities

Summary

This chapter has been concerned with the time planning and control associated with construction projects. The planning of work is not only of interest to clients and their representatives but is also essential to both contractors and subcontractors and will influence the successful completion of a project. In particular, the need for planning and control has been addressed from a practical perspective, in addition to drawing attention to the contractual implications of planning and progressing that contractors may face when tendering and carrying out construction work.

In the planning process itself, the important role of the planner has been discussed, together with the various stages of planning during the construction process. Programming requirements were also outlined in relation to the stages of planning and particular types of construction project. However, in practice much will depend on the parties to the contract and the specific demands and requirements of a project. With regard to the methods of programming, several have been identified. The two most common methods used in practice are still predominantly bar charts and network analysis, and these have been described in more detail.

Planning work by the use of networks initially requires the defining and estimating of durations for activities. Coupled with this, is the requirement of logic and sequencing of activities which essentially represents a thought process carried out by the planner and other team members to establish the most efficient and cost-effective means of carrying out construction work. Current techniques of network analysis were therefore considered in more detail, including arrow diagrams, the precedence convention and the more recent technique of arrodence diagrams. More complex networks, particularly for large projects, often need a different approach, and sectionalised networks and hierarchical subdivision by the use of sub-networks will allow more effective planning and control at different levels of detail.

Monitoring and controlling the progress of a project also represents an essential part of the planning process, and several methods can be adopted to carry out this task efficiently. Without monitoring and control the prepared plans themselves would be of limited value. Additionally, and related to the planning and progressing of work, is the effective management of resources. The efficient utilisation of resources on all projects is often paramount, since they not only have a direct impact on the duration of activities but also influence cost. Time/cost optimisation associated with projects was therefore considered and included compression, complex compression and decompression, which can be applied to a project network.

Many of the techniques used in planning and control are also available through the automated features of application packages. To illustrate the capabilities of such software, SureTrak Project Manager 2.0 from Primavera was used in connection with a small building project. This gave only a brief insight into the power and flexibility of such tools but demonstrated some of the facilities available to planners for planning and controlling construction projects.

References

Ahuja, H. N. *et al.* (1994) *Project Management: Techniques in Planning and Controlling Construction Projects*, second edition, John Wiley & Sons, New York.

Chartered Institute of Building (CIOB), *Construction Computing*, quarterly publication, Ascot, UK.

Construction Industry Computing Association (CICA), *Software Directory*, annual publication, Cambridge, UK.

Cooke, B. and Williams, P. (1998) *Construction Planning, Programming and Control*, Macmillan Press, Basingstoke.

Lester, A. (1991) *Project Planning and Control*, second edition, Butterworth-Heinemann, London.

Lester, A. (1998) 'Arrow diagrams and precedence diagrams: an end to the war?' *Project*, Vol. 11 (1), pp. 21–23, 26.

O'Brien, J. J. (1993) *CPM in Construction Management*, fourth edition, McGraw-Hill, New York.

Oxley, R. and Poskitt, J. (1996) *Management Techniques Applied to the Construction Industry*, fifth edition, Blackwell Science, Oxford.

Paulson, B. C., Jr (1995) *Computer Applications in Construction*, McGraw-Hill.

Pilcher, R. (1992) *Principles of Construction Management*, McGraw-Hill International (UK), London.

Reiss, G. (1995) *Project Management Demystified: Today's Tools and Techniques*, second edition, E & FN Spon, London.

Reiss, G. (1996) *Project Management Demystified: Managing Multiple Projects Successfully*, E & FN Spon, London.

SureTrak Project Manager 2.0 (1997) *User Manual*, Primavera Systems, Bala Cynwyd, Pennsylvania.

Tulip, A. (1998) 'What makes project management software programmes tick', *Project*, Vol. 10 (9) – Vol. 11 (3), London.

3 Financial planning and cost-control systems

Introduction

The following chapter provides details and examples of the attainment of a financial planning and cost-control system. The system is set within the context of a decision-making framework, and the examples are utilised in order to assist in systems implementation.

The decision-making process

Decision making is a task carried out by construction managers on a regular basis. If an organisation is to operate both efficiently and effectively a systematic approach to decision making is required. It is necessary, therefore, for all managers to be able to make effective decisions.

Decisions are required at all organisational levels from the strategic to the operational. The management accountant or quantity surveyor can assist the decision-making processes by indicating to senior management the relative economic advantages of one alternative against another, and this in turn means that decision-making work should concentrate on analysing the economic differences between various options. Figures such as the total cost per unit, total sales or total enterprise profits are of little value here; what is required is the difference in profit that will arise as a result of selecting one alternative in preference to another.

Using this basic approach of analysing, one can consider what will happen if there are differences in activity (break-even charts); if there are major differences in products or fixed costs (cash flow); and if there are differences in the timing of receipts and payments (capital projects – discounted cash flow).

Marginal costing and break-even analysis

Break-even charts are used essentially for making short-term decisions. One must therefore be wary of applying them to long-term or extreme situations, for then much of the theory no longer applies. This warning relates particularly to references to changes in the level of activity.

Fixed and variable costs

Dual basis of costs

Making correct decisions depends very much on understanding how costs behave, and for short-term decisions this in turn depends upon appreciating that all costs are essentially either (1) time-based (i.e. change in proportion to time); or (2) activity-based (i.e. change in proportion to activity).

Time-based (fixed) costs

Time-based costs are costs that change in direct proportion to the length of time that elapses; typical of these is rent. Note that such a cost is not affected by activity. No matter whether the enterprise is busy or slack, the rent payable is the same. Other examples of time-based costs include rates and debenture interest. Since time-based costs do not vary with activity then for any given period of time they will remain unchanged regardless of the level of activity. These costs are therefore termed 'fixed costs'. Note that fixed costs are not costs that never alter; the point being made is that they do not alter as a result of changes in activity.

Activity-based (variable) costs

Activity-based costs are costs that change in proportion to the level of activity undertaken. The most obvious of these costs is direct material costs. If, for example, production is increased by 10%, material costs will also increase. Other such costs are direct wages, sales commission and power. Such costs are termed 'variable costs' and can be defined as costs that vary in direct proportion to activity.

Mixed-based (semi-variable) costs

There are clearly a number of costs that are neither wholly fixed nor wholly variable. Such costs change with activity but not in direct proportion to it. For example, activity may increase by 10%, but a particular cost of this type may increase by only 4 or 5%. Such costs are termed 'semi-variable' and include such costs as maintenance, supervision and store keeping. These costs are really mixed time- and activity-based costs. For instance, maintenance can be analysed into time-based maintenance (e.g. weekly, monthly or annual preventive maintenance, the cost of which is independent of activity) and activity-based maintenance (e.g. 5,000-mile service on car).

Segregation of semi-variable costs

All semi-variable costs can be segregated into their time-based and activity-based (fixed and variable) components. However, in practice it is virtually impossible to segregate the fixed and variable costs by simple inspection (as was suggested with maintenance above); such segregation must be done statistically. The following is a simple method:

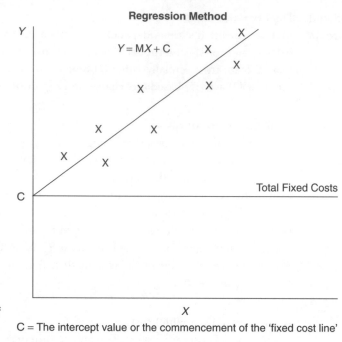

Fig. 3.1 Segregation of semi-variable costs

C = The intercept value or the commencement of the 'fixed cost line'

1. Prepare a graph with axes for activity (horizontal) and costs (vertical).
2. Take the figures from a number of past periods and for each period plot the activity and costs as a single point.
3. Draw the 'line of best fit' through the points and extend it to the cost axis. This is the total cost line.
4. The point where the total cost line cuts the cost axis gives the fixed cost.
5. The variable cost at any level of activity is given by the difference between the fixed cost and the total cost lines.

The object of the above has been to find the regression line of costs on activity. A more accurate method would be to use the regression formula:

$$Y = MX + C$$

where Y = costs and X = activity (see Figure 3.1).

Total costs
Since all semi-variable costs can be segregated into fixed and variable components, all costs in an enterprise can be analysed into fixed and variable categories. This means that the total cost for any period can be regarded as the sum of costs that are wholly fixed and variable costs.

i.e. total costs = fixed + variable costs

Graphing costs and income

It is a simple matter to plot the foregoing equation on a graph having activity and money axes (as indicated in Figure 3.1). On such a graph, the fixed costs will show as a horizontal line at a height equal to the fixed costs, and the total cost curve as a straight upward-sloping line starting from the fixed cost line at nil activity. Since there is nil activity, the variable costs must by definition be nil and the total costs therefore equal to the fixed costs. If a third line, representing revenue (sales) at the different activity levels, is added to the graph, the graph then shows the corresponding profit or loss levels. The sales line will start at the origin of the graph, since nil activity leads to nil revenue (see following example).

The importance of the break-even point

If a company is continually making losses, no matter how small, then its life is definitely limited. On the other hand, if it is making profits, no matter how small, then theoretically it can continue indefinitely. The break-even point is important to management, therefore, since it marks the very lowest level to which an activity can drop without putting the continued life of the company in jeopardy. Occasionally working below the break-even point is not necessarily fatal, but on the whole the company must operate in the long term above this level.

Features of a break-even chart

The following features of break-even charts should be carefully noted:

1. *Costs and income.* The total cost, together with the income, can be read directly off the chart for any chosen level of activity.
2. *Profits and losses.* Since the difference between the total cost of operating and the resulting income is the profit (or loss), then the gap between the total cost line and the income line at any level of activity measures the profit (loss) at that level.
3. *Break-even point.* When the income line crosses the total cost line, income and total costs are equal, and neither a profit nor a loss is made, i.e. the enterprise breaks even. This point is called the break-even point and is usually measured in terms of activity.
4. *Margin of safety.* This is simply the difference between the break-even point and any activity selected for consideration. Figure 3.2 provides a break-even chart example.

The break-even point is measured in terms of activity. Figure 3.2 shows that if the selected activity is 80%, then:

Expected sales	£160,000
Expected total cost	£130,000
Expected profit	£30,000
Margin of safety	20%

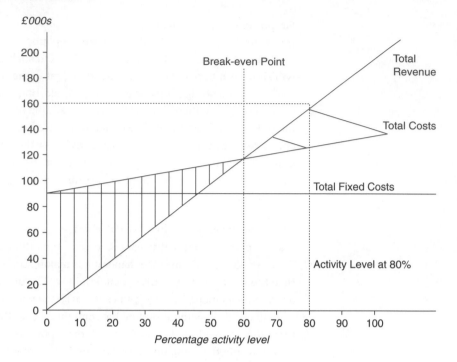

£000s

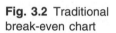

Fig. 3.2 Traditional
break-even chart

Break-even charts are true only within the actual limits of the activity on which they were based (statistically speaking, you cannot extrapolate). To assume that the relationships existing within the range of activities experienced apply outside it is incorrect. If, for instance, the chart was based on activity extremes of 60 and 90%, no attempt should be made to determine figures below 60 or above 90%.

Profit graph

On a profit graph, profit is plotted directly against activity, as per Figure 3.3. At nil activity, therefore, loss equals the fixed costs. Figure 3.3 is based upon the same numerical values as Figure 3.2.

Marginal costing provides a yardstick in the following situations:

1. Deciding prices during a recession.
2. Comparing the results for different contracts.
3. Assessing profits due to increases or decreases in sales.
4. Assessing whether it is suitable to sell below total costs or even the marginal cost, perhaps for a limited period, in order to retain the services of skilled labour during a recession or to maintain production when competition is at its fiercest. There comes a time, however, when costs increase due to inflation, and losses will then become evident.

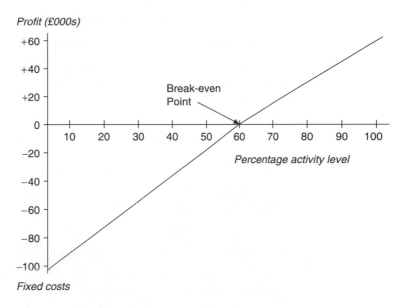

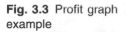

Fig. 3.3 Profit graph example

Figure 3.4 is a pictorial representation of the concepts of marginal costing theory. The following example is provided in the form of questions with advocated solutions.

The example relates to the decision-making process by providing quantitative data.

Questions

A joinery manufacturer currently produces doors. The present capacity of the manufacturer is 5,000 units (doors).

- Associated fixed costs are £100,000.
- Marginal costs are £60 per unit.
- Revenue is £100 per unit.

However, the manufacturer is considering two strategies for expansion:

- The first strategy is to increase output to 10,000 units.
- The second strategy is to increase output to 15,000 units.
- Strategy 1 will entail an increase in fixed costs to £300,000.
- Strategy 2 will entail an increase in fixed costs to £500,000.

1. For the joinery manufacturer, evaluate the three options:
   - no expansion
   - expansion to 10,000 units
   - expansion to 15,000 units.

MARGINAL COSTING BREAKDOWN

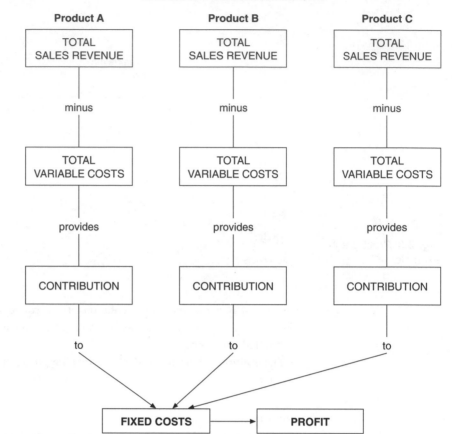

Fig. 3.4 Pictorial representation of the concept of marginal costing

The evaluation is to take the form of a profit graph and should incorporate the break-even points and margins of safety.

2. Check your answer to the graphical evaluation by calculation.
3. Comment upon the evaluation.

Question 1

Option 1 – no expansion.

Fixed costs	£100,000
Marginal costs/unit	£60
Revenue/unit	£100
Total fixed costs	£100,000
Total marginal costs	(£60 × 5,000 units) £300,000
Total revenue	(£100 × 5,000 units) £500,000

total revenue – total marginal costs = contribution
$$£500,000 - £300,000 = £200,000$$

contribution − total fixed costs = profit
£200,000 − £100,000 − £100,000

For Option 1, the profit is £100,000; this equates to £20/unit.

Option 2 − increase production to 10,000 units per annum.

Fixed costs	£300,000
Marginal costs/unit	£60
Revenue/unit	£100
Total fixed costs	£300,000
Total marginal costs	(£60 × 10,000 units) £600,000
Total revenue	(£100 × 10,000 units) £1,000,000

total revenue − total marginal costs = contribution
£1,000,000 − £600,000 = £400,000

contribution − total fixed costs = profit
£400,000 − £300,000 = £100,000

For option 2, the total profit is £100,000; this equates to £10/unit.

Option 3 − increase production to 15,000 units per annum.

Fixed costs	£500,000
Marginal costs/unit	£60
Revenue/unit	£100
Total fixed costs	£500,000
Total marginal costs	(£60 × 15,000 units) £900,000
Total revenue	(£100 × 15,000 units) £1,500,000

total revenue − total marginal costs = contribution
£1,500,000 − £900,000 = £600,000

contribution − total fixed costs = profit
£600,000 − £500,000 = £100,000

For Option 3, the total profit is £100,000; this equates to £6.66/unit. The data are now plotted on a profit graph, as shown in Figure 3.5.

Question 2: checking answers via the calculation method
Option 1 − no expansion.
Calculation of break-even point:

$$\text{break-even point} = \frac{\text{total fixed costs}}{(\text{revenue/unit} - \text{marginal costs/unit})}$$

$$= \frac{£100,000}{(£100 - £60)} = 2,500 \text{ units}$$

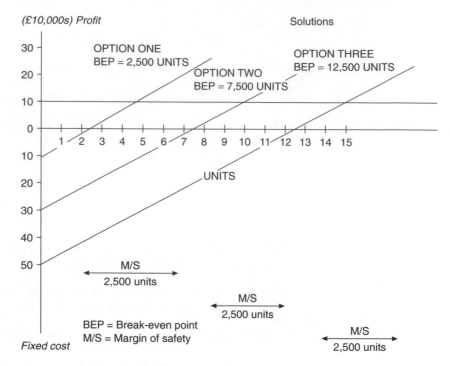

Fig. 3.5 Solutions to the example strategies presented on a profit graph

The break-even point for Option 1 is achieved on the production of 2,500 doors.

Option 2 – expansion to 10,000 units.

$$\text{break-even point} = \frac{£300,000}{£100 - £60} = 7,500 \text{ units}$$

The break-even point for Option 2 is achieved on the production of 7,500 doors.

Option 3 – expansion to 15,000 units.

$$\text{break-even point} = \frac{£500,000}{£100 - £60} = 12,500 \text{ units}$$

The break-even point for Option 3 is achieved on the production of 12,500 doors.

Question 3: commentary on the evaluation

From the evaluation conducted on the three proposed options for this firm, it can be concluded that the increase in production does not increase the profit to be made on the items manufactured. In fact, the margin of profit on each item decreases as the number of units produced increases; this is mostly due to the increase in fixed costs incurred in the production of these items.

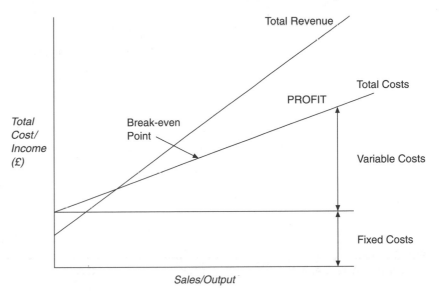

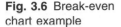

Fig. 3.6 Break-even
chart example

In order to increase the profit margin, it will be necessary for the firm to either reduce the fixed costs or to increase the retail price of its product. The increased output decreases the margin of safety on the break-even point for the items as a percentage of the total output, which can be expressed thus:

Option 1 margin of safety = 95% of overall output

Option 2 margin of safety = 92.5% of overall output

Option 3 margin of safety = 91.7% of overall output

To conclude, expansion of the firm, although it will increase the turnover, will lead to the profit being sacrificed unless the fixed costs can be reduced or the retail price of the goods can be increased to cover the increased cost of manufacture.

The concept of break-even analysis and marginal costing
The most common way of presenting the break-even analysis is in the form of a graph, which provides a medium that is easy to understand and therefore simpler to pass judgment upon. The graph aims to plot the total profit to be made, the total income from the product and the costs in the production of the product identified in Figure 3.6. The cost element in the production of any item or even in the day-to-day running of a business is made up of two main components: fixed costs, i.e. those that change in direct proportion to the passage of time and are not affected by increases in productivity or other outside influences; and variable costs, i.e. those which are affected by changes in productivity such as raw material costs, production, labour, etc. Marginal

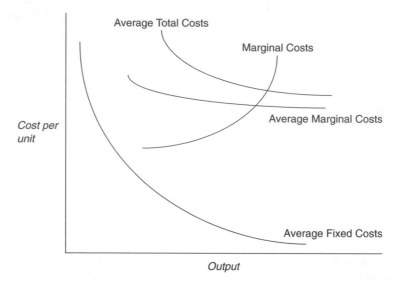

Average Total Costs

Marginal Costs

Cost per
unit

Average Marginal Costs

Average Fixed Costs

Fig. 3.7 Identification
of average costs graph

Output

costs are defined as the cost of producing different quantities of the same
product. This may be expressed as shown in the equation below:

$$\text{marginal cost} = \frac{\text{change in total cost}}{\text{change in quantity produced}}$$

When the average cost of producing an item is decreasing, the marginal cost
for that item must also decrease, and when the cost of producing an item is
increasing so the marginal cost must also be increasing. Thus marginal costing
provides a means of assessing the effects of demand on the price of a product
and the cost of production, therefore giving an indication of the profit margin
on the particular item at any given rate of production.

The concept of marginal costing is explained in the following statement:
once a firm has installed the equipment required for its chosen activity, output, in the short term, is limited to the capacity of the plant, while overheads
such as rent, rates, insurance and staff salaries are fixed. Therefore as production increases the proportion of fixed costs per item produced decreases. This
is true until the plant reaches its capacity and production is at a maximum.
The variable costs per unit such as raw material costs, power and production
labour will initially tend to decrease with output as returns on the variable
factors, but as the plant nears capacity the variable costs will begin to rise
because of the inevitable increase in the costs of production labour; extra
hours means extra money for the operatives. This is presented in Figure 3.7.

Thus marginal costing provides a means of identifying the optimum output
level for a firm and the highest possible profit to be made on that product. The
results of the analysis can be used in a comparative model in order to select the
most profitable use of organisational resources. Remember: 'break-even charts
are only true within the actual limits on which they are based' (Cormican, 1985).

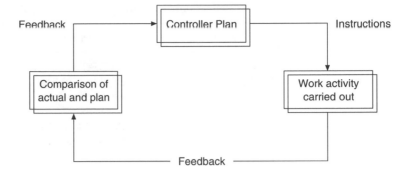

Fig. 3.8 The dynamic control feedback loop

Cash flow analysis and budgetary control systems

Control   Before studying cost control, it is necessary to examine the fundamental principles upon which such control work is based, and this is the objective of this section. The basis of control:

1. *Control fundamentals.* Control is compelling events to conform to a plan, and the essentials of control are:
    *   produce a plan for the activity under consideration;
    *   conduct a comparison of actual results with the planned results;
    *   take action to rectify divergences if necessary.
    If any one of the above is missing, there can be no (effective) control.
2. *The control loop.* All control is dynamic, and this aspect can be depicted in the form of the loop shown in Figure 3.8.
3. *Feedback information.* This enables actual events to be compared with those planned, and on the basis of the divergences revealed the controller can issue fresh instructions.
4. *Control requirements.* For control to be attained, it is necessary that all the elements in the loop are present (and functioning).
5. *Degree of control.* Control may be termed 'tight' or 'loose'. This aspect of control depends upon:
    *   Cycle time – the shorter the control loop cycle time, the tighter the control. This is because the sooner one corrects the divergences the less chance there is of the actual events reaching a point where it will be too late to carry out corrections.
    *   Degree of detail in the explanation of divergences. The more detailed the explanation of divergences, the tighter the control. This is simply because the more informed the controller is regarding divergences, the better his instructions for corrective action. Short cycle times and detailed explanations of divergences are therefore vital for tight control, and managers must bear this in mind at all times.

Table 3.1 School contract example cost/value data

Budgeted costs	Work element	Value of work
45,000	To DPC	50,000
60,000	DPC to first floor	68,000
140,000	1st floor to roof	155,000
40,000	Services	45,000
40,000	Internal finishes	42,000
15,000	External works	17,000
£340,000		£377,000

Budgets   A budget is a cost plan relating to a period of time. Time is a fundamental factor in any budget, a definition that is easily remembered, but it is important to appreciate the complications of the word 'cost'. A cost is the value of an economic resource used. Thus a cost plan is essentially a resource plan in terms of value. It is important to note that it is the quantities which are budgeted, not money figures. The money figures are merely a way of expressing quantities in the form of a common measure.

Cash flows   There is no easy road to the efficient and effective management of cash flows and certainly no one single technique which would guarantee the absence of cash flow problems. The monitoring of cash flows, in particular, requires observation and vigilance. At an individual contract level, the first stage of cash flow management is the relocation of resources and hence cost to each activity. A profit margin can then be produced and a budget developed incorporating costs and expected profit margins. The cash flow monitoring process will then focus on the identification of variances from the projected budget during the actual construction phase. An example has been provided incorporating the development of a cash flow for a project contract and the utilisation of a budgetary control exercise. The example is based around a project scenario.

Cash flow and budgetary control example   A contractor has entered into a contract to build an extension to a school. The terms of the contract are that payment will be made from valuations taken at the end of each month and paid one month later. Retention will be 5%, half paid on practical completion, the remainder six months later. The budgeted costs for the work elements and the value of each element are provided in Table 3.1. The contract programme is as indicated in Figure 3.9. From the information provided below:

- calculate the cash flow for the project;
- plot cumulative cash in and cash out against time.

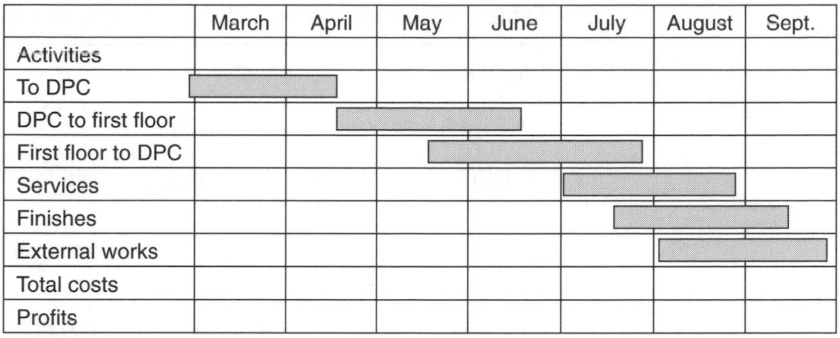

	March	April	May	June	July	August	Sept.
Activities							
To DPC	▓▓▓						
DPC to first floor		▓▓▓▓					
First floor to DPC			▓▓▓▓				
Services					▓▓▓		
Finishes					▓▓▓		
External works						▓▓▓	
Total costs							
Profits							

Fig. 3.9 School contract example contract programme

Table 3.2 Loading the school contract example with costs

Work element	Budgeted costs	Budgeted costs divided by half-monthly units
To DPC	45,000	$\dfrac{45,000}{3} = 15,000$
DPC to first floor	60,000	$\dfrac{60,000}{4} = 15,000$
First floor to roof	140,000	$\dfrac{140,000}{5} = 28,000$
Services	40,000	$\dfrac{40,000}{4} = 10,000$
Internal finishes	40,000	$\dfrac{40,000}{4} = 10,000$
External works	15,000	$\dfrac{15,000}{4} = 3,750$

Table 3.2 is a tabulated representation of the allocation of costs (in half-monthly units) by dividing the cost by the time units. The resulting figures can then be loaded onto the contract programme in order to establish month-by-month costs.

Table 3.3 is a tabulated representation of the allocation of the profit element for each work activity. First, the budgeted costs are subtracted from the value to provide profit. Second, the profit is allocated in the same way as the costs in Table 3.2.

Table 3.4 has the cost and profit elements loaded onto the contract programme. The individual monthly totals have also been established by adding vertically the cost and profit elements encompassed within each month. For example, April has a cost associated with the activity 'to DPC' of £15,000 and a cost associated with the activity 'DPC to first floor' of £15,000. Therefore, the total cost for April is £30,000.

Table 3.3 Loading the school contract example with profits

Work element	Value	Budgeted costs	Profit divided by half-monthly units
To DPC	50,000	45,000	$\dfrac{5,000}{3} = 1,666.70$
DPC to first floor	68,000	60,000	$\dfrac{8,000}{4} = 2,000$
First floor to roof	155,000	140,000	$\dfrac{15,000}{5} = 3,000$
Services	45,000	40,000	$\dfrac{5,000}{4} = 1,250$
Internal finishes	42,000	40,000	$\dfrac{2,000}{4} = 500$
External works	17,000	15,000	$\dfrac{2,000}{4} = 500$

Table 3.4 The school contract example loaded with cost and profit elements

Month / Activities	March	April	May	June	July	August	Sept.	Total
To DPC	30,000	15,000						
	3,334	1,666						
DPC to first floor		15,000	30,000	15,000				
		2,000	4,000	2,000				
First floor to roof			28,000	56,000	56,000			
			3,000	6,000	6,000			
Services					20,000	20,000		
					2,500	2,500		
Finishes					10,000	20,000	10,000	
					500	1,000	500	
External works						7,500	7,500	
						1,000	1,000	
Monthly cost	30,000	30,000	58,000	71,000	86,000	47,500	17,500	340,000
Profit	3,334	3,666	7,000	8,000	9,000	4,500	1,500	37,000

In order to determine the cash flow for the project, the production of a tabulation sheet is most useful. Table 3.5 provides an example format incorporating the school project data. The costs and profits are extracted from the school's contract programme and the loadings on the sheet provide

Table 3.5 Tabulated sheet for the school contract example

CASH FLOW – TABULATED ANALYSIS SHEET

Month	Cost	Profit %	Amount due	Retention 5%	Amount due less retention	Cumulative amount due	Time*-adjusted amount due cumulative	Cumulative costs	Time*-adjusted cumulative costs	Cash flow
1	30,000	3,334	33,334	1,667	31,667	31,667		30,000		
2	30,000	3,666	33,666	1,683	31,983	63,650	31,667	60,000	30,000	+1,667
3	58,000	7,000	65,000	3,250	61,750	125,400	63,650	118,000	60,000	+3,650
4	71,000	8,000	79,000	3,950	75,050	200,450	125,400	189,000	118,000	+7,400
5	86,000	9,000	95,000	4,750	90,250	290,700	200,450	275,000	189,000	+11,450
6	47,500	4,500	52,000	2,600	49,400	340,100	290,700	322,500	275,000	+15,700
7	17,500	1,500	19,000	950	18,050	367,575	340,100	340,000	322,500	+17,600
8				18,850			367,575		340,000	+27,575
9				½ = 9,425						
10										
11										
12										
13						37,700		340,000		
14							377,000		340,000	+37,000
15										

* Graph

Notes:
1. Time-adjusted amount due cumulative column figures provide the stepped lines on the cash flow graph.
2. The time-adjusted cumulative cost figures provide 'S' curve line on the cash flow graph.

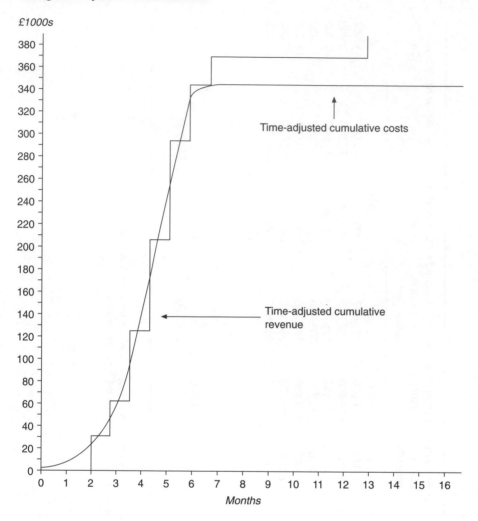

£1000s

Time-adjusted cumulative costs

Time-adjusted cumulative revenue

Months

Fig. 3.10 Standard 'S' curve graph for the school contract example

instructions. These instructions allow the contract conditions to be built into the analysis. For example, the 'amount due' is the extent of the monthly valuation, i.e. costs + profit. The retention column is 5% of the amount due column.

In order to be able to plot the data (see Figure 3.10) and establish a cash flow profile we need to work with cumulative figures. Therefore, the 'cumulative amount due' column is a running total of the 'amount due' column. An allowance should be made for the fact that payments carry a month delay, so we have to adjust the cumulative amount due payment by one month.

Cumulative costs are then established, and they also carry a month delay in payment; this results in the 'time-adjusted cumulative cost' column. Finally, the cash flow column is obtained by subtracting the time-adjusted cumulative costs from the time-adjusted amount due column. In this example, the cash

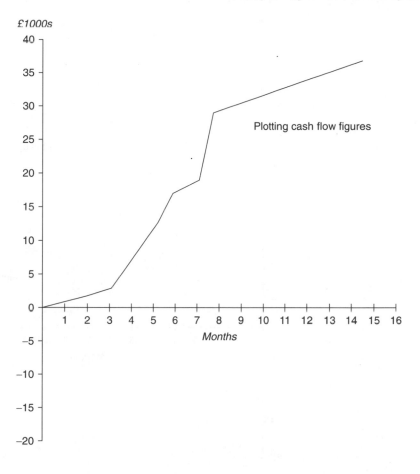

£1000s

Plotting cash flow figures

Months

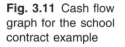

Fig. 3.11 Cash flow graph for the school contract example

flow indicates a cash surplus from month 2 onwards. The cash flow is then plotted as Figure 3.11.

The charts provide a pictorial representation of the projected cash flows for the contract. However, the cumulative nature of cash flow graphs is another vital managerial decision-making information aid.

The cumulative cash flow graph of Figure 3.12 indicates how two (or more) projects can be plotted on the same chart in order to establish the total impact of the dual cash flows on the company. This holistic approach is a vital control mechanism and decision-making aid.

However, Briscoe (1992) opines 'that the time taken to complete construction is a key determinant of cost, as projects which overrun their planned schedules usually incur higher costs than those which are budgeted correctly'.

Having completed the cash flow analysis, it is possible to calculate the interest payments due on the contract. The procedure requires the production of further charts and some calculations. An example follows relating to the school data.

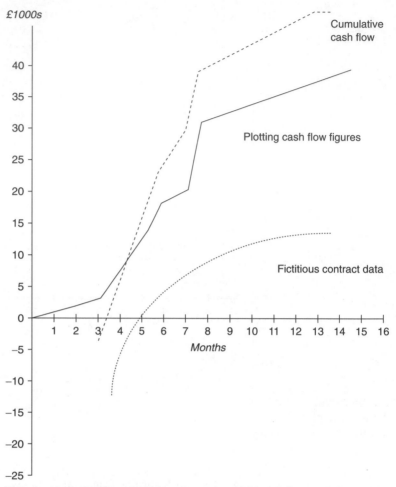

Fig. 3.12 Cumulative cash flow graph example

This chart indicates the cumulative nature of multiple cash flow analysis and its value in establishing the holistic cash flow situation for the host organisation's decision makers

Steps in the calculation of interest due

1. From the tabulated analysis Table 3.5 identify:
   - total adjusted cumulative costs (see Figure 3.13);
   - the difference between costs and revenue (cash flow);
   - then produce Table 3.6.
2. Plot the cash flow graph, Figure 3.14. This includes:
   - negative and positive values from Table 3.6.
3. Produce Table 3.7, i.e. funding required, from Table 3.6.
4. Calculate interest payments as per Table 3.8.

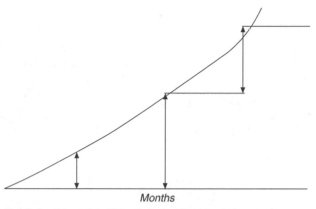

Months

Fig. 3.13 Allocation of cumulative costs for the school contract example

Month 3 will be total at Month 3 – total at Month 2.
Months 1 and 2 are full cash requirements.

Note: The costs are cumulative. Do not subtract anything from Month 1 or Month 2; start at Month 3. This will provide the maximum cash requirement.

Table 3.6 Calculation of interest payments for the school contract example

Month	Time-adjusted cumulative costs	Difference between costs and revenue	Plotting the negative values
1	10,000	0	−10,000
2	30,000	1,667	−30,000
3	60,000	3,650	−30,000
4	118,000	7,400	−58,000
5	189,000	11,450	−71,000
6	275,000	15,700	−86,000
7	322,500	17,600	−47,500
8	340,000	27,575	−17,500
9			
10		For plotting*	
11		positive values	
12			
13			
14	340,000	37,000	0

* 10,000 in Month 1 from cash flow graph; i.e. the cumulative costs read off the 'S' curve graph, Figure 3.10.
Time-adjusted cumulative costs and cash flow are from the tabulated analysis sheet, Table 3.5.
Plotting the negative values: these figures are obtained by subtracting the time-adjusted cumulative costs from each other.

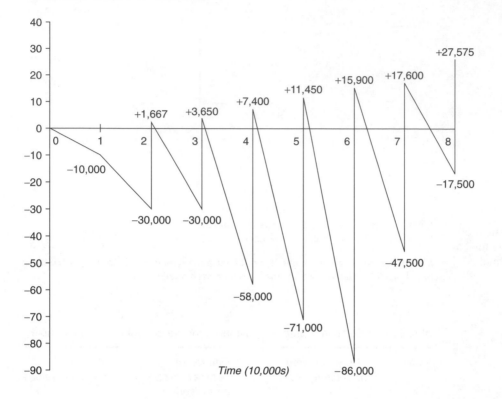

Fig. 3.14 Plotting the negative and positive cash flows for the school contract example

Table 3.7 Calculation of the funding required for the school contract example

Month	Start of month	End of month	Average	Funding period (weeks)
1	0	10,000	5,000	4
2	10,000	30,000	20,000	4
3	0	30,000	15,000	3.75
4	0	58,000	26,000	3.75
5	0	71,000	35,500	3.7
6	0	86,000	43,000	3.5
7	0	47,500	23,750	3
8	0	17,500	8,750	2

Note: £10,000 in Month 1 from cash flow graph.

Budgetary control   The importance of the budgetary control function has previously been established. Now one can apply the theory to the school project example. However, we should note that the main task is to break down the contract expenditure into its various categories (maybe elements of structure) with a view to allocating money (i.e. budgets) against each for the duration of the contract. This

Table 3.8 Calculation of the average monetary sum lock-up for the school contract example

Month	£	Weeks	Interest at 20%
1	5,000	4	$\dfrac{5,000 \times 0.2 \times 4}{52} = 76.92$
2	20,000	4	$\dfrac{20,000 \times 0.2 \times 4}{52} = 307.69$
3	15,000	3.75	$\dfrac{15,000 \times 0.2 \times 3.75}{52} = 21.63$
4	26,000	3.75	$\dfrac{26,000 \times 0.2 \times 3.75}{52} = 375$
5	35,500	3.7	$\dfrac{35,500 \times 0.2 \times 3.7}{52} = 505.19$
6	43,000	3.5	$\dfrac{43,000 \times 0.2 \times 3.5}{52} = 578.84$
7	23,750	3	$\dfrac{23,750 \times 0.2 \times 3}{52} = 274.03$
8	8,750	2	$\dfrac{8,750 \times 0.2 \times 2}{52} = 67.30$
Total			£2,206.60

Note: Total interest charges of £2,206.60 will be due utilising the cash flow profile established.

Table 3.9 Breakdown of elemental costings for the school contract example

Activities	Labour £	Materials £	Plant £	Site overheads £	Totals £
To DPC	20,000	10,000	14,500	500	45,000
DPC to first floor	30,000	25,000	4,200	800	60,000
First floor to roof	50,000	80,000	9,000	1,000	140,000
Services	12,000	26,000	1,600	400	40,000
Finishes	18,000	21,500	–	500	40,000
External works	7,000	2,000	5,500	500	15,000
Total	137,000	164,500	34,800	3,700	340,000

has been done in Table 3.9. Budgeted operating costs are the most realistic expectation of what actual costs will be. Actual operating costs are the most realistic expectation of what actual costs will be. Actual costs may be compared with budgeted costs (at budgeted output) as an overall budget check on

Table 3.10 Percentage completions incorporated matrix for the school contract example

Activities	% Completed	Labour	Materials	Plant	Site o/h	Totals
To DPC	100	20,000	10,000	14,500	500	45,000
DPC to first floor	100	30,000	25,000	4,200	800	60,000
First floor to roof	70	35,000	56,000	6,300	700	98,000
Services	48	5,760	12,480	768	192	19,200
Finishes	30	5,400	6,450	–	150	12,000
Ext. works	15	1,050	300	825	75	2,250
Totals £	–	97,210	110,230	26,593	2,417	236,450

the efficiency of the project. A breakdown of the budgeted cost for each activity for the School Project example is provided below:

The basis of budgetary control activity is the comparison of actual costs with budgeted costs (as advocated in the dynamic control loop, see page 75). Therefore, we require some actual costs for comparison; these have been provided for the end of July on the programme. The percentage completions for the end of July are also tabulated.

Actual costs at the end of Month 5 (July):

Labour	£101,000
Materials	£115,200
Plant	£ 27,500
Site O/H	£2,500

Percentage completions (end of July):

To DPC	100%
DPC to first floor	100%
First floor to roof	70%
Services	48%
Finishes	30%
External Work	15%

The first step in the analysis is to produce a table incorporating the percentage completions. This is Table 3.10, which contains the adjustment of Table 3.9 values in accordance with the percentage of work actually completed.

Upon the completion of Table 3.9 a variance analysis table can be produced as Table 3.11. This required the adjusted budgeted costs from Table 3.10 to be compared with the stated actual costs. Actual costs are subtracted from adjusted budgeted costs. The variance figures can then be used to calculate the percentage variance on the contract, as indicated in Table 3.11.

Table 3.11 Variance analysis table for the school contract example

Cost centre	Budgeted cost	Actual cost	Variance	
			+	−
Labour	97,210	101,000		3,790
Materials	110,230	115,200		4,970
Plant	26,593	27,500		907
Site o/h	2,417	2,500		83
Totals	236,450	246,200		−9,750

Table 3.12 Individual variance calculations per cost centre matrix for the school contract example

Cost centre	Variance on cost centre from previous table	Individual variance $\dfrac{(Variance)}{\left(\begin{array}{c}Total\ adjusted\ budget\\ for\ cost\ centre\end{array}\right)} \times 100$
Labour	−3,790	$\dfrac{(-3,790)}{(97,210)} \times 100 = -3.899\%$
Materials	−4,970	$\dfrac{(-4,970)}{(110,230)} \times 100 = -4.508\%$
Plant	−907	$\dfrac{(-907)}{(26,593)} \times 100 = -3.410\%$
Site o/h	−83	$\dfrac{(-83)}{(2,417)} \times 100 = -3.434\%$
Totals	−9,750	

Total variance as a percentage of total budgeted costs:

$$\% \text{ variance} = \frac{\text{Total variance}}{\text{Total budgeted costs}} \times 100$$

$$= \frac{-9,750}{236,450} \times 100$$

Therefore, variance on budgeted costs is −4.12%.

The usefulness of variance analysis depends upon the timeliness of the activity and the detail of the analysis. The activity needs to be conducted at regular intervals – remember that one cannot have retrospective control. Knowing the total variance is not all that useful for managers in the decision-making process; it would be more useful to know the individual variance of each cost element. These can be calculated as shown in Table 3.12.

These individual figures are of greater value to the contracts manager than the overall figure. They provide details of individual cost centre performances. It should be noted that the analysis can be worked with the cost centres as the elements of the structure rather than labour and plant, etc. If an analysis is conducted utilising both approaches to cost centres, a comprehensive analysis is provided generating valuable data for control activities.

Capital investment appraisal

In this section, we examine decision-making methods that relate to long-term projects, i.e. projects running into years and particularly capital projects. When evaluating long-term projects, the fixed/variable analysis becomes less important as the emphasis is on the project cost rather than the cost per unit. On the other hand, the timing of cash receipts and payments often becomes critical. Davis *et al.* (1996) advocate that 'improvements in efficiency and output will be related to the quantity of investment being undertaken'.

Three methods of investment appraisal

Surveys suggest that construction companies employ a variety of methods for investment appraisal. Approximately 90% utilise quantitative methods to evaluate investment, based on three main methods. The most popular is the 'payback' method, which assesses a project in terms of how soon a business can recoup its investment. However, some companies use more sophisticated calculations based on 'rates of return' and 'cash flow', often in combination.

Payback method
This refers to the length of time which elapses before the initial outlay on an investment is recovered out of the stream of profit generated by the project. Other things being equal, an investment project which pays for itself quickly will be preferred to one which takes longer.

Average annual percentage rate of return
This method measures the anticipated stream of profit to be earned over the expected lifetime of a project, expressed as an overall annual percentage rate. An investment project is viable when its internal rate of return exceeds the rate of interest, which measures the 'cost' of financing the investment.

Net present value
This method measures the anticipated stream of net cash flows over the expected lifetime of a project, expressed in terms of their 'discounted' present value. The greater an investment project's net present value, the more viable it will be compared with other projects.

Table 3.13 Data for appraisal techniques example

Year	Cash flow £	Cumulative cash flow £
1	5,000	5,000
2	4,000	9,000
3	9,000	18,000 AMOUNT TO RECOVER ←
4	6,000	24,000
5	3,000	27,000

Internal rate of return

This method also discounts a stream of cash flows and is a means of establishing the true rate of return on a project after the cash flows have been discounted. This rate can then be compared with the project's cut-off rate.

Application of the three appraisal techniques

The following provides an example of the application of all the above techniques of investment appraisal. A simple project cash flow has been generated for the application of the techniques, and the data are shown in Table 3.13. Capital invested is £20,000.

Method one

The amount to be recovered is £20,000. After three years we have recovered £18,000; therefore a shortfall of £2,000 exists. The next increment is £6,000, therefore

$$\frac{\text{shortfall}}{\text{next increment}} \times 12 = \text{time in months to recover shortfall}$$

$$\text{Therefore } \frac{2,000}{6,000} \times 12 = 4 \text{ months}$$

Pay back period for project = 3 years 4 months

Method two

Average annual percentage rate of return (AA%RR).
Step 1: calculate the average annual rate of return (AARR).

$$\text{AARR} = \frac{\text{total return (see Table 3.14)}}{\text{investment period}}$$

$$= \frac{27,000}{5}$$

$$= £5,400$$

Table 3.14 Average annual percentage rate of return

Year	Cash flow £
1	5,000
2	4,000
3	9,000
4	6,000
5	3,000
Total	27,000

Table 3.15 Tabulated NPV data at 14%

Year	Cash flow £	NPV.F (14%)	NPV £
0	−20,000	1.00	−20,000
1	5,000	0.88	4,400
2	4,000	0.77	3,080
3	9,000	0.67	6,030
4	6,000	0.59	3,540
5	3,000	0.52	1,560
Total			−1,390

Step 2: Calculate AA%RR.

$$AA\%RR = \frac{AARR}{capital\ invested} \times 100$$

$$= \frac{5,400}{20,000} \times 100$$

$$= 27\%$$

Method three

Based upon the inverse of compound interest.

$$compound\ interest = (1 + i)n$$

$$discounted\ cash\ flow = \frac{1}{(1 + i)n}$$

$$= net\ present\ value$$

Let discount rate (cut-off rate) be 14%. Table 3.15 indicates that we do not receive sufficient funds to warrant the expenditure economically. The return on capital expenditure is not equal to the £20,000 invested and the compound interest due at a cut-off rate of 14%.

Table 3.16 Tabulated NPV data at 10% and 12%

Year	Cash flow £	NPV.F (12%)	NPV £	NPV.F (10%)	NPV £
1	5,000	0.89	4,450	0.91	4,550
2	4,000	0.80	3,200	0.83	3,320
3	9,000	0.71	6,390	0.75	6,750
4	6,000	0.64	3,840	0.68	4,080
5	3,000	0.57	1,710	0.62	1,860
Totals			19,560		20,560
			−410		+560

From the previous example (NPV), it has been established that we are not obtaining a return equal to 14%. We can therefore ask the question, 'what is the true rate of return?' The return is not as high as 14%, so let us try 12% first. Table 3.16 establishes that at 10% NPV.F the project makes a profit of £560 and at 12% NPV.F the project makes a loss of £410. The true NPV can be obtained by interpolation as follows:

The IRR is between 12 and 10%. Let us use interpolation to obtain the IRR:

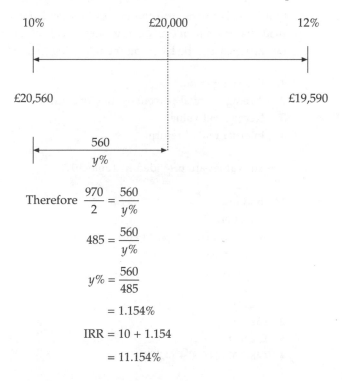

Therefore $\dfrac{970}{2} = \dfrac{560}{y\%}$

$485 = \dfrac{560}{y\%}$

$y\% = \dfrac{560}{485}$

$= 1.154\%$

$IRR = 10 + 1.154$

$= 11.154\%$

The techniques have been applied to one project. However, their main advantage is as a decision-making aid when evaluating alternative options. Therefore, the following case study and solution provides a fuller example of their use.

Investment appraisal example

A construction organisation has £80,000 in its capital expenditure budget. The managing director has invited proposals for the investment of funds. Two proposals have been received.

The first bid was from the Plant Department, which proposed buying four new fuel-efficient lorries as replacements for 8-year-old ones. Each would cost £20,000 and should last 4 years, after which time they would have a second-hand value of £4,000. The new lorries should save £8,000 each per year on fuel and maintenance costs.

The only other detailed proposal came from the Marketing Department, which requested the establishment of a new customer services department. It would cost £80,000 to set up, and its £20,000 per year running costs should generate an extra contribution from sales amounting to:

Year 1	£30,000
Year 2	£40,000
Year 3	£80,000
Year 4	£80,000

Our brief is to produce the necessary calculations to enable a choice to be made based on financial analysis only. A cut-off rate of 10% is to be used, and our analysis is to be based on the following methods of evaluation:

1. Payback period.
2. Average annual percentage rate of return.
3. Net present value.
4. Internal rate of return.

Present values are provided in Table 3.17.

Method one
Payback period.
    Project A (Plant Department).
    Capital invested: £80,000.
    Income (years):

1   £32,000
2   £32,000
3   £32,000
4   £48,000 (32,000 + (4,000 × 4))

The amount to be recovered is £80,000. After 2 years, £64,000 will have been recovered, as shown in Table 3.18; therefore, a shortfall of £16,000 exists. The next increment will be £32,000. Therefore:

Table 3.17 Net present value tables for examples

Rate / Years	1%	2%	3%	4%	5%	6%	7%	8%	9%	10%	11%	12%
1	0.99	0.98	0.97	0.96	0.95	0.94	0.93	0.93	0.92	0.91	0.90	0.89
2	0.98	0.96	0.94	0.92	0.91	0.89	0.87	0.86	0.84	0.83	0.81	0.80
3	0.97	0.94	0.92	0.89	0.86	0.84	0.82	0.79	0.77	0.75	0.73	0.71
4	0.96	0.92	0.89	0.85	0.82	0.79	0.76	0.74	0.71	0.68	0.66	0.64
5	0.95	0.91	0.86	0.82	0.78	0.75	0.71	0.68	0.65	0.62	0.59	0.57
6	0.94	0.89	0.84	0.79	0.75	0.70	0.67	0.63	0.60	0.56	0.53	0.51
7	0.93	0.87	0.81	0.76	0.71	0.67	0.62	0.58	0.55	0.51	0.48	0.45
8	0.92	0.85	0.79	0.73	0.68	0.63	0.58	0.54	0.50	0.47	0.43	0.40
9	0.91	0.84	0.77	0.70	0.64	0.59	0.54	0.50	0.46	0.42	0.39	0.36
10	0.91	0.82	0.74	0.68	0.61	0.56	0.51	0.46	0.42	0.39	0.35	0.32
11	0.90	0.80	0.72	0.65	0.58	0.53	0.48	0.43	0.39	0.35	0.32	0.29
12	0.89	0.79	0.70	0.62	0.56	0.50	0.44	0.40	0.36	0.32	0.29	0.26
13	0.88	0.77	0.68	0.60	0.53	0.47	0.41	0.37	0.33	0.29	0.26	0.23
14	0.87	0.76	0.66	0.58	0.51	0.44	0.39	0.34	0.30	0.26	0.23	0.20
15	0.86	0.74	0.64	0.56	0.48	0.42	0.36	0.32	0.27	0.24	0.21	0.18

Rate / Years	13%	14%	15%	16%	17%	18%	19%	20%	30%	40%	50%
1	0.88	0.88	0.87	0.86	0.85	0.85	0.84	0.83	0.77	0.71	0.67
2	0.78	0.77	0.76	0.74	0.73	0.72	0.71	0.69	0.59	0.51	0.44
3	0.69	0.67	0.66	0.64	0.62	0.61	0.59	0.58	0.46	0.36	0.30
4	0.61	0.59	0.57	0.55	0.53	0.52	0.50	0.48	0.35	0.26	0.20
5	0.54	0.52	0.50	0.48	0.46	0.44	0.41	0.40	0.27	0.19	0.13
6	0.48	0.46	0.43	0.41	0.39	0.37	0.35	0.33	0.21	0.13	0.09
7	0.43	0.40	0.38	0.35	0.33	0.31	0.30	0.28	0.16	0.09	0.06
8	0.38	0.35	0.33	0.31	0.28	0.27	0.25	0.23	0.12	0.07	0.04
9	0.33	0.31	0.28	0.26	0.24	0.23	0.21	0.19	0.09	0.05	0.03
10	0.29	0.27	0.25	0.23	0.21	0.19	0.18	0.16	0.07	0.03	0.02
11	0.26	0.24	0.21	0.20	0.18	0.16	0.15	0.13	0.06	0.02	0.01
12	0.23	0.21	0.19	0.17	0.15	0.14	0.12	0.11	0.04	0.02	0.008
13	0.20	0.18	0.16	0.15	0.13	0.12	0.10	0.09	0.03	0.013	0.005
14	0.18	0.16	0.14	0.13	0.11	0.10	0.09	0.08	0.03	0.009	0.003
15	0.16	0.14	0.12	0.11	0.09	0.08	0.07	0.06	0.02	0.006	0.002

Table 3.18 Tabulated data for Project 'A': payback method

Year	Cash flow £	Cumulative cash flow £
1	32,000	32,000
2	32,000	64,000 AMOUNT TO RECOVER
3	32,000	96,000
4	48,000	144,000

Table 3.19 Tabulated data for Project 'B': payback method

Year	Cash flow £	Cumulative cash flow £
1	10,000	10,000
2	20,000	30,000 AMOUNT TO RECOVER
3	60,000	90,000
4	60,000	150,000

$$\frac{\text{shortfall}}{\text{next increment}} \times 12 = \text{time in months to recover shortfall}$$

$$\frac{16,000}{32,000} \times 12 = 6 \text{ months}$$

Payback period for project A = 2 years 6 months

Project 'B' (Marketing Department).
  Capital invested: £80,000:
  Income (years):

1   £10,000
2   £20,000
3   £60,000
4   £60,000

The amount to be recovered is £80,000. After 2 years, £30,000 will have been recovered, as shown in Table 3.19; therefore, a shortfall of £50,000 exists. The next increment is £60,000. Therefore:

$$\frac{\text{shortfall}}{\text{next increment}} \times 12 = \text{time in months to recover shortfall}$$

$$\frac{50,000}{60,000} \times 12 = 10 \text{ months}$$

Payback period for project B = 2 years 10 months

Table 3.20 Tabulated data for Project 'A': average annual percentage rate of return method

Year	Cash flow £
1	32,000
2	32,000
3	32,000
4	48,000
Total	144,000

Method two

Average annual percentage rate of return (AA%RR).

Project A (Plant Department).

Step one: calculate the average annual rate of return (AARR):

$$AARR = \frac{\text{total return (see Table 3.20)}}{\text{investment period}}$$

$$= \frac{£144,000}{4}$$

$$= £36,000$$

Step two: calculate the AA%RR:

$$AA\%RR = \frac{AARR}{\text{capital invested}} \times 100$$

$$= \frac{£36,000}{£80,000} \times 100$$

$$= 45\%$$

Project B (Marketing Department).

Step one: calculate the average annual rate of return (AARR):

$$AARR = \frac{\text{total return (see Table 3.21)}}{\text{investment period}}$$

$$= \frac{£150,000}{4}$$

$$= £37,500$$

Table 3.21 Tabulated data for Project 'B': average annual percentage rate of return method

Year	Cash flow £
1	10,000
2	20,000
3	60,000
4	60,000
Total	150,000

Table 3.22 Data for Project 'A': net present value method

Year	Cash flow £	NPV.F (10%)	NPV £
0	−80,000	1.00	−80,000
1	32,000	0.91	29,120
2	32,000	0.83	26,560
3	32,000	0.75	24,000
4	48,000	0.68	32,640
Total			32,320

Step two: calculate the AA%RR:

$$AA\%RR = \frac{AARR}{\text{capital invested}} \times 100$$

$$= \frac{£37,500}{£80,000} \times 100$$

$$= 46.88\%$$

Method three

Based upon the inverse of compound interest:

compound interest $= (1 + i)n$

$$\text{discount cash flow} = \frac{1}{(1 + i)n}$$

Discount cash flow: net present value
Project A (Plant Department).
Discount rate (cut-off rate) is 10%. Table 3.22 establishes that sufficient funds are received to warrant the expenditure, and an overall profit of £33,320 will be made.

Table 3.23 Data for Project 'B': net present value method

Year	Cash flow £	NPV.F (10%)	NPV £
0	−80,000	1.00	−80,000
1	10,000	0.91	9,100
2	20,000	0.83	16,600
3	60,000	0.75	45,000
4	60,000	0.68	40,800
Total			31,500

Table 3.24 Data for Project 'A': internal rate of return method

Year	Cash flow £	NPV.F (20%)	NPV £	NPV.F (30%)	NPV £
1	32,000	0.83	26,560	0.77	24,640
2	32,000	0.69	22,080	0.59	18,880
3	32,000	0.58	18,560	0.46	14,720
4	48,000	0.48	23,040	0.35	16,800
Totals			90,240		75,040
			+10,240		−4,960

Project B (Marketing Department).

Discount rate (cut-off rate) is 10%. Table 3.23 establishes that sufficient funds are received to warrant the expenditure, and an overall profit of £31,500 will be made.

Method four

Project A (Plant Department).

From the previous methods, it has been established that any return will not be equal to 10%. So what is the true rate of return?

The return is higher than 10%, so try 20%. Table 3.24 establishes that at 20% a profit of £10,240 will be made and at 30% a loss of £4,960 will be made. Interpolation can be used to establish the internal rate of return.

The IRR is between 20 and 30%. Use interpolation to obtain the IRR:

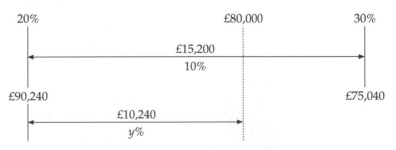

Table 3.25 Data for Project 'B': internal rate of return method

Year	Cash flow £	NPV.F (20%)	NPV £	NPV.F (30%)	NPV £
1	10,000	0.83	8,300	0.77	7,700
2	20,000	0.69	13,800	0.59	11,800
3	60,000	0.58	34,800	0.46	27,600
4	60,000	0.48	28,800	0.35	21,000
Totals			85,700		68,100
			+5,700		−11,900

Therefore $\dfrac{£15,200}{10} = \dfrac{£10,240}{y\%}$

$$£1,520 = \dfrac{£10,240}{y\%}$$

$$y\% = \dfrac{£10,240}{£1,520}$$

$$= 6.74$$

$$IRR = 20\% + 6.74\%$$

$$= 26.74\%$$

Project B (Marketing Department).

From the previous methods, it has been established that any return will not be equal to 10%. So what is the true rate of return?

The return is higher than 10%, so try 20%. Table 3.25 establishes that at 20% a profit of £5,700 will be made and at 30% a loss of £11,900 will be made. We can now use interpolation to calculate the internal rate of return.

The IRR is between 20 and 30%. Use interpolation to obtain the IRR.

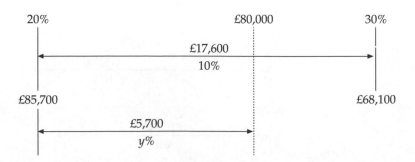

Table 3.26 Data for Project 'A': payback period including net present value

Year	Cash flow £	NPV.F (10%)	NPV	Cumulative cash flow £
1	32,000	0.91	29,120	29,120
2	32,000	0.83	26,560	55,680
3	32,000	0.75	24,000	79,680
4	48,000	0.68	32,640	112,320

$$\text{Therefore} \quad \frac{£17,600}{10} = \frac{£5,700}{y\%}$$

$$£1,760 = \frac{£5,700}{y\%}$$

$$y\% = \frac{£5,700}{£1,760}$$

$$= 3.24$$

$$\text{IRR} = 20\% + 3.24\%$$

$$= 23.24\%$$

The following uses the payback period after the application of discounting the funds flow.

Method five

Payback period including net present value.

Project A (Plant Department).

The amount to be recovered is £80,000. After 3 years, £79,680 will have been recovered, as shown in Table 3.26; therefore, a shortfall of £320 exists. The next increment is £48,000. Therefore:

$$\frac{\text{shortfall}}{\text{next increment}} \times 12 = \text{time in months to recover shortfall}$$

$$\frac{320}{48,000} \times 12 = 0.007 \text{ months}$$

Payback period for project A = 3 years

Project B (Marketing Department).

The amount to be recovered is £80,000. After 3 years, £70,700 will have been recovered; therefore, a shortfall of £9,300 exists as identified in Table 3.27. The next increment is £60,000. Therefore:

Table 3.27 Data for Project 'B': payback period including net present value

Year	Cash flow £	NPV.F (10%)	NPV	Cumulative cash flow £
1	10,000	0.91	9,100	9,100
2	20,000	0.83	16,600	25,700
3	60,000	0.75	45,000	70,700
				←
4	60,000	0.68	40,800	111,500

Table 3.28 Matrix analysis: comparing results for Project 'A' and Project 'B'

Technique	Project results	Ranking
Pay-back period	Plant Department 2 years 6 months	1st
	Marketing Department 2 years 10 months	2nd
Average annual percentage rate of return	Marketing Department 46.88%	1st
	Plant Department 45%	2nd
Net present value at 10%	Plant Department £32,320	1st
	Marketing Department £31,500	2nd
IRR	Plant Department 26.74%	1st
	Marketing Department 23.24%	2nd

$$\frac{\text{shortfall}}{\text{next increment}} \times 12 = \text{time in months to recover shortfall}$$

$$\frac{9,300}{60,000} \times 12 = 2 \text{ months}$$

Payback period for project A = 3 years 2 months

Matrix analysis sheet

For ease of decision making, the analysis can be presented in a matrix format, as indicated in Table 3.28. As can be seen from the table, all but the average annual percentage rate of return method place the Plant Department option first. However, senior management may have some qualitative element that also requires consideration.

Summary

In this chapter, a number of different decision-making techniques have been outlined, and the reader may well be asking how can one tell which technique should be selected for a specific problem. There is, unfortunately, no simple answer, but the following should help a selection to be made with a certain amount of confidence.

First, it must be emphasised that the fundamental idea in decision-making work is to examine differences between alternatives – or between undertaking a project and not undertaking it. Application is not always easy; however, the reader should introduce into any analysis every factor that differs between alternatives and which will affect the profit of the enterprise (e.g. changes in fixed costs; changes in contribution and/or interest charges). It is permissible to ignore every factor that remains the same whichever alternative is selected.

Beyond this there are no firm rules; ultimately, the only guide is a sound appreciation of what is required in the circumstances surrounding the whole of the decision-making situation. The various decision-making techniques should be viewed as a means of providing a holistic system for controlling the finances of a construction-related organisation.

References

Briscoe, G. (1992) *The Economics of the Construction Industry*, B. T. Batsford, London.

Cormican, D. (1985) *Construction Management Planning and Finance*, Longman, Harlow, England.

Davis, B., Hale, G., Smith, C., and Tiller, H. (1996) *Investigating Economics*, Macmillan, London.

4 Quality systems and performance

Introduction

The development of quality assurance systems as aids to management has been in progress for many years, with the focus on applications pertaining to manufacturing industries. In a construction environment, however, this is still somewhat new. Although the construction process can be compared with the manufacturing process, the design and production of a building differs in many ways from the design and manufacture of products.

Some of the essential differences of quality in building are many and varied. Almost all construction projects are 'unique', with the building process representing a single production run. The tradition in construction has been to separate the design and construction processes, while manufacturing industry adopts an integrated approach. The construction site is 'individual' in terms of its temporary environment. The life-cycle of a construction project from inception to completion extends beyond the manufacturing cycle and also tends to evolve and develop through time. The considerable mobility of construction staff prevents the development of long-term production teams, and each construction site is likely to have different team members.

Feedback from the building in use to the designer is remote from the actual time of design and construction and often prevents the effective analysis of defective design and construction, whereas in manufacturing, testing for deficiency and necessary corrective actions can be implemented quickly. The management of quality in the design and construction process is essential to ensure the required quality of service.

The value of quality assurance in construction

Essentially, there is a requirement to provide an assurance that design and construction aspects have the capability to produce a product that is effective and economic, whether that product is the design of the building or the construction of the building. The pursuit of quality commences with the client and continues through the production process to the utilisation of the building. Quality assurance is therefore an integral part of the 'total building process'.

Quality is in many ways subjective and a matter of judgment. To provide a clear view of the meaning of quality, Griffith (1990) defines a number of aspects which should be considered:

Function: does the building meet the requirement? Life: is the building durable? Economy: does the building represent value for money? Aesthetics: is the building pleasing in appearance and compatible with its surroundings? Depreciation: is the building an investment?

The interpretation and measurement of quality are as ambiguous as its perception. Clients will have their own idea of the quality required to meet stated needs and desires.

The architect's aim is to provide value for money, assisted by the quantity surveyor and bounded by an acceptable standard of construction. On site, the quality of work is dependent on the skills and application of the craft operatives, or 'workmanship'. It can therefore be hypothesised that one's view of quality is dependent upon one's involvement and role in the overall construction process. Quality in construction can therefore be determined by expectations. Dalton (1988) emphasises that

management of quality and quality itself are closely related to a number of various expectations surrounding the performance of buildings, these being quality, durability and reliability.

Quality assurance systems

Quality assurance is concerned with planning and developing the technical and managerial competence needed to achieve the host organisation's desired objectives. Quality assurance is also concerned with the management of people, addressing the roles, duties and responsibilities of individuals within the organisation. Quality assurance is primarily the responsibility of management; its structure and implementation must become part of the total organisational framework. Quality assurance must also be an important aspect of the marketing and promotional strategy of the organisation. Only when quality assurance pervades the entire organisation and becomes an integral and recognised aspect of its operations will it foster the potential to become truly successful in giving the organisation a competitive advantage.

Quality assurance must be actively employed throughout the total building process from initial briefing and conceptual design, through the assembly process to the completion of the project. It is essential that clear communication is encouraged, in particular at the interfaces of project responsibility and control.

Quality assurance is concerned with developing a 'formal' structure, organisation and operational procedures to ensure specified quality throughout the total building process. The construction industry can be divided into five broad sectors where quality assurance is applicable:

1. *Client* in the production of the project brief.
2. *Designer* in the design and specification process.
3. *Manufacturers* in the supply of materials, products and components.
4. *Contractors* (and *subcontractors*) in construction, supervision and management processes.
5. *User* in the utilisation of the new structure.

There are few standards and codes that affect the client and the final procurement and use of the building, the majority of quality assurance applications being present in the manufacturing sector of the construction industry.

CIRIA (1989) highlights the responsibilities towards quality assurance:

Quality cannot be inspected into a product or project; it must be built in. Responsibility lies with those doing the work – the client, the architect, the engineer, the contractor, the operatives, the materials suppliers and the sub-contractors.

BS/EN/ISO 9000 series quality systems provide a framework for the five sectors noted above. Successful implementation provides a system which has obtained certification by the recognised body. This in turn provides monitoring and enforcement authority to an independent third party.

The BS/EN/ISO 9000 series provides a certificated structure with which all systems seeking certification must comply. The framework for quality assurance has become fully developed and operational in Europe. In summary, the essential features of the BS/EN/ISO 9000 series are:

1. The appointment of a senior manager (generally known as the quality manager) in the organisation, who is responsible for the quality system.
2. A well-documented system of procedures and instructions.
3. Records of all inspections and audits.
4. Adequate training of all staff.
5. Segregation of rejected products, materials and documents/drawings so that they cannot be used by mistake.
6. Adequate packaging for delivery.

PSA/DoE (1986) note that 'Only when Quality Assurance encompasses all stages of a building's life i.e. design, construction, final evaluation and maintenance will maximum benefit be achieved.'

Summary   The British Standards Institution (BSI) is the largest and most important certification organisation currently involved in the development of quality assurance systems, having led the way towards improving quality standards throughout the 1960s and the 1970s, and publishing BS 4891: 'A Guide to Quality Assurance' in 1972 and 'A National Strategy for Quality' in 1978. The BS/EN/ISO 9000 series is now the UK's national standard for all new quality assurance certification schemes. The introduction of quality assurance systems provides an organisational framework to attain the required level of quality consistently.

BS/EN/ISO 9000 series quality assurance systems

Any company wishing to become registered for quality assurance must satisfy the requirements of a 'certification body', and these certifiers in turn must be approved or accredited by government. A certification body is defined as an impartial body, government or non-governmental, possessing the necessary competence and reliability to operate a certification scheme and in which the interests of all parties concerned with operational aspects of the system are represented. The term 'accreditation' is defined as the formal recognition by a national government, against published criteria, of the technical competence and impartiality of a certification body or testing laboratory. Third-party certification is conducted by an independent body which has no contractual relationship with the client and/or contractor. The purpose of third-party certification is to provide confidence that products or services supplied comply with the specified requirements. There are a number of third-party quality assurance schemes currently in operation with specific relevance to the building and civil engineering industries. However, the most prominent one is the BS/EN/ISO 9000 series quality system. The certification process can be represented in the following three aspects of implementation.

Examination of documentation   Applicants for third-party certification must submit evidence that proves they have a documented quality system complying with the requirements of the standard. Documentation will normally consist of a 'quality manual', a 'procedures manual' and, if required, a set of 'work instructions'. Should any discrepancies exist between the requirements and company documentation, the company will be notified of requested amendments by the certification body.

Assessment   Once the documentation has been accepted by the certifying body, it will perform an on-the-spot investigation (audit) in order to verify that the documentation system is being implemented as described. If everything is satisfactory, then a certificate will be awarded. The certificate may relate to the whole company or to specific activities stated on the certificate.

Monitoring   After the initial assessment, regular visits are made (usually two to four per annum) to ensure that standards are being maintained. If standards are not being maintained, the ultimate sanction would be the withdrawal of certification.

The route to certification is depicted in Figure 4.1.

Cost implications of the certification process

Potential benefits from establishing and maintaining a certified quality assurance system are not secured without costs to the organisation. These costs are both direct and indirect. Significant direct costs are incurred in

- developing the quality assurance system;
- producing the quality documentation;
- establishing the implementation system;
- maintaining the internal audit system; and
- independent third-party assessment.

Indirect costs are difficult to assess but can include

- liaising with the certification body;
- changes to operational processes and procedures to accommodate certification requirements;
- some demotivational aspects associated with staff and the implementational process; and
- the consumption of organisational energy and efforts during the drive for certification.

There are also the costs of maintaining the system and surveillance visits by the certification body. Certification bodies specify their various registration fees, which are subject to some variation depending on the following factors:

- size of the company and number of employees;
- structure of the organisation;
- diversity and range of the company's activities;
- nature and complexity of the quality system; and
- complexity of the documentation.

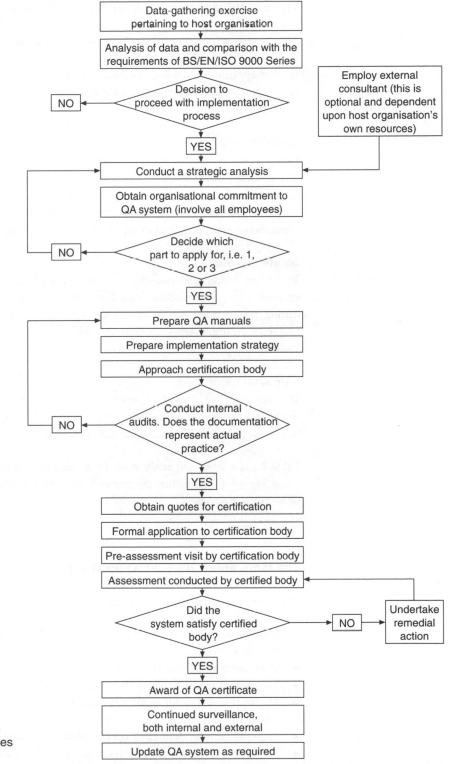

Fig. 4.1 The route to BS/EN/ISO 9000 series certification

The constituent parts of BS/EN/ISO 9000 series quality systems

The above noted quality standards are divided into four constituent parts, with 1–3 providing three levels of quality assurance. Part 1 is the most stringent level of application, while Parts 2 and 3 relate to lower orders of severity pertaining to system requirements, and Part 4 provides guidance on the application of Parts 1–3.

BS/EN/ISO 9001

Part 1 systems contain all 20 clauses of the quality system and are therefore the most comprehensive system available for certification. Organisations involved in the design process and seeking certification need to comply with Part 1 requirements. For example, this specification would apply to design consultants, architects, and design and build companies.

BS/EN/ISO 9002

This is applicable to manufacturing and/or installation companies that do not undertake design activities. Part 2 requires process and inspection control during production and/or service delivery. This standard applies to most construction organisations, where evidence of inspection and tests, during a construction process, has to be provided to clients.

BS/EN/ISO 9003

This applies to organisations where quality assurance systems can be based upon final inspection of product or service only. This part of the standard has limited application in the construction industry.

Table 4.1 is a tabulated analysis of the applicable clauses for Parts 1, 2 and 3, while Figure 4.2 identifies the contents of both the quality manual and the procedures manual.

The facts about BS/EN/ISO 9000 series quality systems

The following points summarise the essential points of the quality system:

- The requirements of the quality system can be adapted to suit individual organisations.
- Once designed, the system can be subject to change and development.
- The system is not bureaucratic in its application.
- The responsibility for quality is shared fairly with those who control the actual work.
- The organisation defines its own quality policy and objectives.
- The organisation must continually assess and review its quality system.

Table 4.1 BS/EN/ISO 9000 series quality system elements

Subclause numbers in BS/EN/ISO 9004	Title	Corresponding subclause numbers		
		BS/EN/ISO 9001	BS/EN/ISO 9002	BS/EN/ISO 9003
4.1	Management Responsibility	4.1	4.1	4.1
4.2	Quality System	4.2	4.2	4.2
4.3	Contract Review	4.3	4.3	
4.4	Design Control	4.4		
4.5	Document Control	4.5	4.4	4.3
4.6	Purchasing	4.6	4.5	
4.7	Purchaser Supplied Product	4.7	4.6	
4.8	Product Identification and Traceability	4.8	4.7	4.4
4.9	Process Control	4.9	4.8	
4.10	Inspection and Testing	4.10	4.9	4.5
4.11	Inspection, Measuring and Test Equipment	4.11	4.10	4.6
4.12	Inspection Status	4.12	4.11	4.7
4.13	Control of Non-conforming Product	4.13	4.12	4.8
4.14	Corrective Action	4.14	4.13	
4.15	Handling, Storage, Packaging and Delivery	4.15	4.14	4.9
4.16	Quality Records	4.16	4.15	4.10
4.17	Internal Quality Audits	4.17	4.16	
4.18	Training	4.18	4.17	4.11
4.19	Servicing	4.19		
4.20	Statistical Techniques	4.20	4.18	4.12

Quality Manual

GENERAL INFORMATION ON A COMPANY
POLICY STATEMENT
ORGANISATION
MANAGEMENT REVIEW
QUALITY SYSTEM

Procedures Manual

CONTRACT REVIEW
DESIGN CONTROL
DOCUMENT CONTROL
PURCHASING
PURCHASER SUPPLIED PRODUCT
PRODUCT IDENTIFICATION AND TRACEABILITY
PROCESS CONTROL
INSPECTION AND TESTING
INSPECTION, MEASURING AND TEST EQUIPMENT
INSPECTION AND TEST STATUS
CONTROL OF NON-CONFORMING PRODUCT
CORRECTIVE ACTION
HANDLING, STORAGE, PACKAGING AND DELIVERY
QUALITY RECORDS
INTERNAL QUALITY AUDITS
TRAINING
SERVICING
STATISTICAL TECHNIQUES

Fig. 4.2 Contents of the quality and procedures manuals

Documentation

Quality manual

The quality manual provides an overview of the organisation and incorporates a declaration by the chief executive that the company is committed to enforcing the documented procedures. The scope and field of application for the procedures are also identified. A history of the company and services offered must be included, accompanied by a short statement pertaining to each clause of the relevant quality assurance system stating simply that the company intends to operate the procedure. This is the document organisations use for marketing purposes.

Procedures manual

A standard approach to the production of procedures is vital to the successful outcome of the certification process. The use of standard pro formas in the writing of procedures is advocated. Consideration should be given to developing procedures in four interlinked stages, encompassing purpose, scope, responsibility and procedure.

Purpose

This explicitly defines what the procedure intends to achieve; for example, the 'contract review procedure' is concerned with ensuring that all customer requirements are fully understood. The organisation also needs to state that it will ensure that it has the necessary resources and facilities to meet all the requirements of the contract. Any differences or non-conformance with the specified requirements must be discussed with the client, and this has to be noted in the purpose section of the procedure. Finally, the purpose needs to make clear that adequate records of all aspects of the procedure will be retained.

Scope

The scope must establish the areas to which the procedure applies. For the area of contract review, the scope should state that the procedure covers the receipt of all enquiries and orders for the supply of construction-related products/services.

Responsibility

This aspect of the procedure establishes who will perform tasks associated with the purpose and scope. Relating the responsibility section to the previous contract review procedure example, something similar to the following would be incorporated. The project manager is responsible for liaising with the customer throughout the contract review stages, for verification of customer and supplier capability, and for preparation of the tender.

Procedure

This should identify exactly the processes to be followed by all engaged in the operational aspects of a procedure. It must be prescriptive, so that no

Best Construction

BEST CONSTRUCTION
PROCEDURAL STANDARD FORM
FOR CONTRACT REVIEW.

SECTION 3.0 OF A QUALITY
SYSTEM TO MEET THE REQUIREMENTS
OF BS/EN/ISO 9001

Revision Number	Revision Details	Date	Approved By
0	First Issue		

Fig. 4.3 Procedural pro forma front sheet

employee is left in any doubt about the methodology to be adopted. Should further information be necessary to support the procedure, 'work instructions' can be written for staff. These documents supplement the procedure and are really fourth-tier documents.

All procedures are written using the above approach and when collated form the procedures manual. Example pro formas are provided in Figures 4.3 (procedural pro forma front sheet) and 4.4, 4.5 and 4.6 (procedure format sheets).

	Section: 3.0
Construction	Sheet: 1 of
Quality Procedures	Issue No: 1
Manual	Revision No: 0
Subject: Contract Review	

Procedure No.	
3.0	Purpose
3.0.1	To ensure that the requirements of the client are fully understood by Best Construction.
3.0.2	To ensure that Best Construction has the necessary resources and organisational facilities to conform to all of the requirements in the contract.
3.0.3	To ensure that any differences or nonconformity to the specified requirements are discussed with the client and resolved.
3.0.4	To ensure that adequate records of the above are made and retained.

Issued By: Quality & Facilitator	Date of Issue:	Approved By:

Fig. 4.4 Pro forma procedure format (purpose)

	Section: 3.0
Construction	Sheet: 2 of
Quality Procedures	Issue No: 1
Manual	Revision No: 0
Subject: Contract Review	

Procedure No.	
3.1	Scope
	This procedure covers the receipt of all enquiries and orders for the supply of construction-related activities.
3.2	Responsibility
3.2.1	The enquiry receiver is responsible for obtaining enquiry details from the client and documenting in line with this procedure. All tender documentation shall be recorded.

Procedure No.	
3.2.2	The contracts manager is responsible for appointing a team to process the enquiry.
3.2.3	The contracts manager is responsible for liaison with the client throughout the contract review stages, for verification of customer and supplier capability, and preparation of the tender.
3.2.4	The managing director is responsible for authorising acceptance of contracts.

Issued By: Quality & Facilitator	Date of Issue:	Approved By:

Fig. 4.5 Pro forma procedure format (scope and responsibility)

	Section: 3.0
Construction	Sheet: 4 of
Quality Procedures	Issue No: 1
Manual	Revision No: 0
Subject: Contract Review	

Procedure No. 3.3	Procedure Note This section of the procedure would document the exact operational activities of the company pertaining to the contract review section.
Issued By: Quality & Facilitator	Date of Issue:	Approved By:

Fig. 4.6 Pro forma procedure format (procedure)

Testing the theoretical advocated advantages of implementation

This section summarises the findings of a research project pertaining to the attainment of the theoretical advocated advantages of certification to BS/EN/ISO 9000 series quality standards. The theoretical advantages have been established and correlated with the actual results of construction firms operating a certificated quality system. Field research was conducted via structured questionnaires. In summary, it can be stated that most of the theoretical advocated advantages do exist in a construction-related operational environment.

Many construction-related enterprises have now achieved certification to BS/EN/ISO 9000 series. These certificated organisations have endured the pains of certification in the hope that certain benefits can be obtained. However, this section establishes the theoretical benefits and engages in formulating evidence of their existence in reality.

Rationale for the utilisation of BS/EN/ISO 9000 series quality systems

Over half a century ago, an eminent author stated the following:

What is the problem of control of quality of manufactured products? To answer this question, let us put ourselves in the position of a manufacturer turning out millions of the same kind of thing every year whether it be lead pencils, chewing gum, bars of soap, telephones or automobiles, the problem is the same. He sets up a standard for the quality of a given kind of product. He then tries to make all pieces of product conform with this standard. Here his troubles begin. For him quality is a bull's-eye, he often misses.

(Shewhart, 1931)

Although the above quotation was written in 1931, it is just as pertinent today. What is required is some kind of system that will engender consistency throughout the organisation. This consistency of hitting the bull's-eye, i.e. providing what the customer requires, will lead to continued customer satisfaction. The system which was required back in 1931 is in fact a 'quality assurance system'. The BS/EN/ISO 9000 series provides the requirements of such a quality system. A quality assurance system helps companies to:

- focus clearly on the needs of their markets;
- achieve a top-quality performance in all areas, not just in production or service quality;
- operate the simple procedures necessary for the achievement of a stated quality performance;
- critically and continually examine all processes to remove non-productive activities and waste;
- see the improvements required and develop measures of performance attainment;
- understand fully and in detail its competition, and develop an effective competitive strategy;
- develop a team approach to problem solving;
- develop good procedures for communication and the acknowledgment of good work;
- continually review the processes to develop the strategy of never-ending improvement.

(Oakland, 1990)

Today's business environment is such that managers must strive for competitive advantage to hold on to their market share, let alone increase it; price is no longer the major determining factor in customer choice; it has been replaced by quality. Roche (1980) identified from his research studies that 'quality was the first choice for 37% of respondents with price second at 27%'.

In a highly competitive world, customers are becoming more quality-conscious. They know that the bitterness of poor quality lingers on long after the sweetness of low cost has been forgotten. To hold on to customers, makers of products and service providers require a quality plan. This will directly define their efforts towards customer satisfaction. The BS/EN/ISO 9000 series is unique as a set of standards because it does not deal with a particular product but rather assesses the quality system as a whole. It is related to the company's effectiveness in delivering a quality product or service to the customer from the contractual and design phase to installation, storage and maintenance. The standard recognises that a company cannot function smoothly and prove that it has been doing so unless there are clearly kept records of what has been done and what procedures should be followed. Employees cannot be expected to perform at the optimum level if they are never told precisely what they should be doing and there are no guidelines on their specific duties.

BS/EN/ISO 9000 series specifications may on the face of it seem demanding; however, thousands of companies are now certificated to them. Therefore, the

effort expended in obtaining the award must be offset by the benefits of certification.

> The direct benefits to firms who have been assessed in relation to BS 5750
> and who appear in the Department of Trade and Industry's Register of
> Quality Assessed United Kingdom Companies are considerable: reduced
> inspection costs, improved quality and better use of scarce resources. Export-
> ing firms who have been assessed will find that assessment helps them
> to obtain reciprocal recognition of Certificates where needed by overseas
> authorities. (BSI 1987)

The above is important when one considers that there is increasing demand in
many overseas markets for certification of industrial and consumers goods.

Another advocated advantage is that 'As more British Standards become
harmonised with international ones, B.S.I. Certification will be an increasing
help to you in export markets' (BSI, 1989). Companies have also developed a
greater understanding of the importance and real meaning of quality over the
last few years. They realise that the monitoring of quality is not a dreary
routine but rather a proactive way to achieve greater customer satisfaction
and a higher profit margin. As a result, companies have accepted the need for
an effective quality system (Pick-Up, 1990).

The above are the advocated theoretical advantages as documented by
various eminent institutions and authors. However, research has been conducted into the purported advocated advantages by asking 100 registered
construction companies why they sought and obtained certification to BS/
EN/ISO 9000 standards. The following is a summary of the theoretical advocated advantages:

1. Provides a marketing focus.
2. Provides a means of achieving a top-quality performance in all areas of
   the organisation.
3. Provides operating procedures for all staff.
4. Critical audits are performed, allowing for the removal of non-productive
   activities and the elimination of waste.
5. Provides a quality advantage as a competitive weapon.
6. Develops group/team spirit within the company.
7. Improvement of communication systems within an organisation.
8. Reduces inspection costs.
9. More efficient utilisation of scarce resources.
10. Recognition of certification overseas.
11. Customer satisfaction, i.e. provides the required customer quality every
    time.
12. At the request of your customer(s) (reason for certification).

Theoretical advocated advantages of certification to BS/EN/ISO 9000 series	Response to questions					Advantage proved		
	Greatly	Hardly	Not at all	YES	NO	To a great extent	Partly	Not to any great extent
1. Provides a marketing focus	72	24	4			*		
2. Provides a means of achieving a top-quality performance in all areas	76	17	7			*		
3. Provides operating procedures for all staff				Q1 34	66	*		
				Q2 88	12			
4. Audits remove non-productive activities and eliminate waste	Q1 28	53	19					
	Q2			37	63			*
5. Provides a competitive weapon	Q1 70	19	11			*		
	Q2			82	18			
6. Develops group/team spirit	26	55	19					*
7. Improves communications	43	49	8				*	
8. Reduces inspection costs	12	45	43					*
9. More efficient utilisation of resources	Q1 9	53	28					*
	Q2 12	46	42					
	Q3 15	53	32					
	Q4 29	45	26					
	Q5 29	51	30					
10. Provides recognition overseas	Q3 5	23	72	Q1 79	21			*
				Q2 94	6			
11. Provides the required customer quality every time	Q1 50	40	10					
				Q2 53	47		*	

KEY: Q = question

Fig. 4.7 Analysis of field research pertaining to theoretical advocated advantages of implementing BS/EN/ISO 9000 series

Figure 4.7 provides an analysis of the results of a completed questionnaire and conclusions drawn therefrom.

Conclusions

Theoretical advocated advantages relating to efficiency

The noted theoretical advantages associated with an increase in the efficiency of the company when the BS/EN/ISO 9000 series is implemented have not been proved to any great extent.

The advantages that fall under the efficiency heading are:

1. that the series enables the removal of non-productive activities and eliminates waste;
2. that the series reduces the costs associated with inspection;
3. the series provides for the efficient use of resources pertaining to
    * manpower
    * money
    * machines
    * management
    * materials.

The failure of the series to provide the theoretical advocated advantages is attributable to the inherent nature of the standard. The BS/EN/ISO 9000 series is designed to allow a company to assure customers that it can provide for their requirements. The assurance is that the company can provide a product/service that is fit for its purpose. In order to assure the customer of its capability, the company produces procedures. These procedures are documented, and the documented procedures are then audited. The purpose of the audit is to make sure that what the company states as its operating procedures are in fact being implemented.

However, what the company states in the documentation may not in fact be the most efficient process(es) to be employed in order to produce the product/services required by the customer. The BS/EN/ISO 9000 series is concerned with effectiveness, i.e. achieving one's objectives (providing customer satisfaction). It is not designed to provide a sudden increase in organisational efficiency. Therefore, it can be stated that establishments considering using the BS/EN/ISO 9000 series as a means of obtaining a sudden increase in the operating efficiency of their organisations should reconsider their motives. Incorporating the BS/EN/ISO 9000 series quality system into their operational activities is unlikely to provide the increased efficiency anticipated.

Theoretical advocated advantages relating to competitive advantage

The competitive advantage concept of the BS/EN/ISO 9000 series has been proved in practice to a great extent by field research. Establishments seeking to gain a competitive advantage via certification to the series would appear to be able to attain their objective.

Theoretical advocated advantages relating to communications and customer requirements

Both improved communications and the satisfaction of customer requirements have been only partially proved by field research. The results indicate that in both cases it was proved in approximately 50% of the sample. It would appear that implementation of the series is not a guarantee of improved communications or that customer requirements can be satisfied on every occasion.

Disadvantages of certification to BS/EN/ISO 9000 series standards

The research project sought to identify any disadvantages that organisations had found by implementing the series. The responses are indicated in Figure 4.8. However, it is worth noting that 54 companies reported no disadvantages at all.

Summary Before one states whether the theoretical advocated advantages are or are not obtainable in practice, perhaps one should consider the reasons why construction companies seek certification to the series. From the research conducted, the main reasons identified by construction companies for wanting BS/EN/ISO 9000 series quality system certification were established as

Number of responses	10	20	30	40	50	60	(R)
1. No disadvantages at all							54
2. Public image: i.e., high company profile							1
3. Time-consuming procedures							1
4. Persuading work force to accept responsibility for quality							2
5. Cost of implementation							10
6. Requirement of extra manpower to implement							1
7. Cost of being registered							9
8. Disorganised BSI office							1
9. Quantity of extra paperwork							6
10. Unnecessary customer audits							4
11. Exposes company limitations							3
12. Some large companies ignore ISO 9000 and require their own standard							4
13. Increased management bureaucracy							1
14. Not accepted by some government departments as equal to A. QAPT, 4 to 9							3

Fig. 4.8 Noted disadvantages of implementing BS/EN/ ISO 9000 series

(R) = response rate

- provides a means of obtaining a top-quality performance for the company;
- provides a competitive weapon;
- provides a means of ensuring customer satisfaction; and
- provides a means of establishing operational procedures for all staff.

The advantages that can be obtained in practice have been identified. However, when drawing conclusions reference needs to be made to the reasons that companies deemed to be the most important advantages. Considering the responses of the construction companies, it can be stated that the most important advocated advantages can be obtained in practice. Certainly, the most important ones from the sampled companies' point of view are that

- The series provides a marketing focus for the company.
- The series provides a means of achieving a top-quality performance in all areas of the host organisation.

- Implementation leads to a far greater understanding of organisational procedures by all employees.
- The series provides a competitive advantage for the host organisation.

It is therefore possible to conclude that most of the important theoretical advantages do exist in practice.

Problems associated with implementation

This section focuses on the practical problems associated with implementing a BS/EN/ISO 9000 series quality system. The findings are based upon a comprehensive literature review and field research conducted on 100 construction-related firms, all certificated under the series. The previous section has established that the main advocated advantages of implementation do exist in practice. However, the implementation process can be a most problematic one. The following establishes the problems and suggests solutions designed to ease the implementation process for construction-related organisations. Senior management support is the most vital element in a successful implementation process. If senior management support is not forthcoming, the quality facilitator/manager (the person charged with the implementation of the quality system) could also face further problems, such as:

- a lack of adequate authority
- insufficient funding for the project
- a lack of sufficient time allocated for the project
- resistance to
  - documentation gathering
  - implementation during the project.

A successful implementation process is dependent upon the strong commitment and involvement of the senior management of the host organisation. That commitment also needs to be demonstrated through 'policies' and 'support'. If organisations are to avoid problems pertaining to resource issues, senior management must provide the necessary resources. The two most important resource issues are those of adequate funding for the project and allowing sufficient time for people to participate. Participation is necessary when the quality facilitator is gathering information to write the quality and procedures manuals. The participation of staff is also vital during the implementation phase of the project. It should be noted that time allocation and funding are not mutually exclusive. A lack of funds can mean that money is not available to release staff when participation is requested. Issues of authority and overcoming resistance to change are also not mutually exclusive. 'If appropriate authority does not accompany managerial responsibilities and duties, the manager's effectiveness within the organisation is impaired' (Glassman, 1978).

Glassman suggests that managers be delegated sufficient authority to complete their allocated tasks. Senior management needs to ensure that middle managers are not asked to perform tasks for which they have not the necessary authority. There may well be some resistance to change within the host organisation. Coalitions of resistance could develop, and if they are linked to a power base they could impede the implementation process. It is worth noting the differing strategies that could be adopted for the implementation process and the likely outcome of each. In order to do this, an overview of modernist and postmodernist organisational theory is required.

A brief comparison of modernist and postmodernist assumptions

Modernist theory assumes that change is a linear process and therefore can be managed in an incremental way with distinct points of conception and completion. In essence, it is a belief in a simple cause–effect relationship; in such a world, it is easy to achieve any desired outcomes. However, a more realistic view of the operational environments of business organisations rejects the notion of linearity.

Postmodernist organisations realise that change can go in many directions, and the world is best understood in terms of disorder and unpredictability. If one accepts the postmodernists' view, one must also recognise the need for versatility of approach and the emphasis must be placed on flexibility. This emphasis on flexibility must focus on the complexity of boundary relationships and heterarchy as opposed to hierarchy. Another vital consideration is the acceptance of ambiguity by the host organisation.

Modernism *vs* postmodernism
In times of static or limited dynamic environmental change, the modernist (bureaucratic) organisational structure can cope with the change process reasonably well. But when the operational environment becomes dynamic and complex, the structured modernist organisation finds it difficult to cope with the implications of change management.

As Passmore (1994) states:

Most of us are born with a good deal of flexibility; it's a helpful trait that allows our species to adapt to the wide range of habits and circumstances we encounter. But the process of growing up in a hierarchical world teaches us to become inflexible.

Passmore is therefore advocating that people can inherently deal with change (however, they may not like it), and it is the bureaucratic systems they work in that stifle their inherent flexibility.

Some of the early authors upon this subject, such as Weber (1908), suggested that

> Modern business enterprises are structured as 'rational-legal' hierarchical and bureaucratic systems characterised by standardised operating procedures, regulations, performance standards and 'rational' decision-making processes that are based upon technical and professional expertise.

This is now being contested by various authors. Two such are Morris and Brandon (1993), who suggest that there has been a paradigm shift in the way organisations view themselves and their operational environments. After all,

> When the business world undergoes change, only those companies that react quickly will prosper. This ability to react requires considerable flexibility and an openness to new ideas and approaches. In creating this foundation the basic assumptions of the business must be re-examined.

This paradigm shift has manifested itself in the postmodernist organisation, in which employees are better suited to change.

The structure of relationships

Within modernist organisations, very simple structural or boundary relationships exist. Linkages are achieved through formal rules and procedures, and relationships between different groups are also formalised. In comparison with this, the postmodernist organisation possesses little distinctiveness of roles and boundaries are blurred. There exists a greater emphasis upon creating teams and positive, productive relationships, all directed at increasing the organisation's ability to cope with a dynamic environment. This is necessary if the organisation is to be creative. Majaro (1992) points out that making this change to a postmodernist organisation 'is easier said than done' and that 'one of the most difficult challenges to any organisation is the process of changing a climate or corporate attitudes'. It is undoubtedly a difficult change process for an organisation to undergo; however, the benefits are well worth the efforts.

Hierarchy

The modernist organisation has a very defined hierarchy with defined leadership roles. These roles are fixed by legitimacy and tradition. There are leaders and followers.

Contrasted with the modernist organisation is the postmodernist organisation, where a normal hierarchy does not exist and staff act according to agreed areas of expertise. The term for this approach is 'heterarchy', in which very high levels of fluidity exist. This high level of fluidity is a basic necessity, because 'Too much is changing for anyone to be complacent' (Peters, 1988). Within the postmodernist organisation, each task may have its own mini-hierarchy, depending on the needs of the situation. As organisations move to areas of increased complexity of service, there is a requirement to implement increasingly heterarchic ways of working.

Mechanistic *vs* holistic

In the modernist organisation, the relationships between tasks are of a mechanistic nature. There is also a high degree of linearity between organisational tasks. Within the postmodernist organisation, however, high levels of group work exist, each with a correspondingly high level of autonomy. The overriding linking force binding these empowered groups together is the organisational culture. This form more readily suits the reality of today's environment, because organisations and markets are messy things and not linear. One must not forget that building a shared culture and conception of the world takes a great deal of time and effort. It is our view that culture is the 'DNA' of organisations, and this must be 'genetically' engineered to provide the organisation required. Traditionally, in most organisations the existing culture is based upon mistrust and the use of frequent sanctions by senior managers.

Determinacy *vs* indeterminacy

The modernist organisation conducts all matters in a deterministic manner. There is a high degree of emphasis on imposed stability, control and discipline. This assumes that one can exercise a high degree of control over the operational environment. However,

> Many companies feel the 'hot breath of change' in their necks . . . They need to successfully change their organisations into more productive and innovative ones.
>
> (Vander Erve, 1993)

In the postmodernist organisation, matters are conducted in a way that emphasises indeterminacy. This is an acknowledgment that the environment is highly unpredictable and uncertain. This kind of organisation values different things to the modernist organisation; for example, flexibility and innovation are highly prized:

> Flexible people are open-minded, willing to take reasonable risks, self-confident, concerned and interested in learning. They are creative and willing to experiment with new behaviours in order to make better choices about what works for them and the organisation . . . They possess basic skills that

allow them to adapt readily to new circumstances, and they view themselves as able to make the best of opportunities that come their way.

(Passmore, 1994)

This, in essence, is the postmodernist organisation.

Causality

The major difference between modernist and postmodernist organisations upon this issue is that modernists view causality as being linear. They view every element of organisational life as having a cause-and-effect relationship and consequently they manage the organisation in this light.

The postmodernist, when considering causality, thinks of a circle. That is to say, they are encouraged to look for complexity and the interconnectedness of cause and effect. This demands a high level of staff participation, which makes good management sense. The rationale for participation has been stated by Sayles (1989): 'When subordinates are consulted about and contribute to the change process many benefits accrue.'

Morphostatic *vs* morphogenic

Morphostatic processes are defined as those that support or preserve the present mode of operation. These include formal and informal control systems. The emphasis in this type of organisation is upon formal control systems and procedures. A more enlightened approach is adopted by the postmodernist organisation, where a morphogenic culture exists. Morphogenic processes are those that tend to allow for change and development and thus empower the development of quality assurance systems. The exciting nature of change is always advocated. This type of organisation allows staff to be proactive and not reactive, a vital aspect of quality assurance.

All the above identified characteristics of the postmodernist company are essential for any organisation to be able to operate both efficiently and effectively in a dynamic and turbulent operational environment. Organisations require variety in their approach, and hierarchical, authoritarian organisations are poorly equipped to provide such variety. Only business organisations based upon the postmodernist paradigm with vastly reduced bureaucratic control and a rich array of horizontal communication channels, where workers are given a substantial share of power to make choices and to develop new ideas, can survive under new market conditions. This is a fundamental requirement if the responsibility for the quality of products and services is to be rightly placed with the workforce. The quality facilitator should try to overcome resistance by allaying employees' fear of change. Many an implementational process has failed due to persistence with an outmoded organisational structure. Senior management must effect a truly morphogenic change and not a

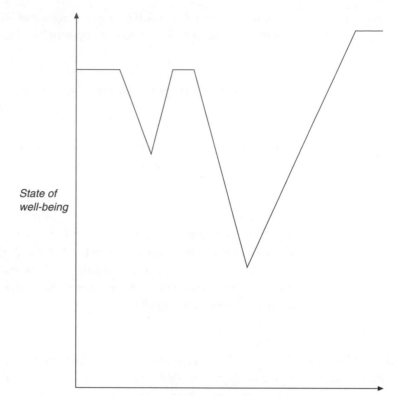

State of well-being

Time

Fig. 4.9
Implementational
change model

cosmetic, morphostatic one. Senior management has to take an active role in both designing and implementing the quality system. Senior management support for the system must be in evidence from the start of the project. This commitment can be shown through the development of organisational policies which involve them in designing and implementing the quality system. Managers within the host organisation are obliged to 'manage'. They should not abdicate to the quality facilitator (team) without providing adequate authority. Liebmann (1993) offers corroboration by noting that when he was part of a team implementing a quality system he found that senior managers 'Charged as to design a process to empower employees but did not empower the team.' The result was failure of the 'quality project'.

Even before the implementation process begins all staff need to be made aware of the benefits of certification to the ISO9000/EN29000/BS5750 series. Staff members need to be convinced that the introduction of such a quality system is worth while and can provide advantages for them and the host organisation. It is therefore senior management's duty to echo the rationale for and the advantages of certification. This is an important implementational point, because people have a built-in resistance to change. Figure 4.9 underlines the importance of obtaining the co-operation of all staff.

This model indicates that the state of well-being among employees is influenced by the implementation of change. A small drop in the state of well-being occurs when employees are informed that change is going to take place. However, as is normal in organisations, after a short period of time they feel that it will not happen and they return to their original state of well-being. When the change process is implemented, employees realise that change really is going to take place and their state of well-being falls. How far this drop in morale is allowed to continue depends to a great extent upon senior management. If, as is advocated, all staff are kept informed of the rationale for and advantages of the change, the fall can be a short one. It is obvious that the further that staff morale is allowed to fall the greater the task of lifting it back to a satisfactory state. In fact, as the model shows, it is possible to have a higher state of well-being after a successful implementation process.

The co-operation of staff is vital for the successful implementation of the quality system. In order for staff to co-operate, two issues require consideration:

1. Staff must want to be co-operative.
2. Staff must be allowed to co-operate.

If staff are not coerced into being co-operative they will provide a greater contribution to the implementation process. This issue is linked very closely to the morale issue.

It can be concluded that senior management support is a vital component at all stages in the design and implementation process. If this support is not provided, a successful outcome to the project can be put in doubt. Whoever is charged with the task of designing and implementing a quality system to BS/EN/ISO 9000 series they must have the total support of the host organisation. This total support includes not only senior management but also the employees (the people who perform the documented tasks). If the implementers can obtain this total support for the system then a successful implementation can be achieved. An important part of obtaining total involvement is informing people what the system is all about and keeping them informed throughout the design and implementation process. Successful implementation of any quality model depends most strongly upon the people who are involved in it. Lozier and Teeter (1993) postulate that

> To improve processes requires teams of people whether these teams are confined to an organisational unit or they represent several cross-functional units. These teams must understand the process, and they must create an atmosphere in which people feel comfortable, confident, motivated and responsible for conducting their work.

The organisational model advocated by Lozier and Teeter is in fact the postmodernist paradigm.

Generic overview of staff project roles within a quality assurance system

The following provides an overview of the roles played by on-site staff members; however, as noted in the heading to this section, only a generic approach is possible.

Site manager

The site manager carries the burden of responsibility for ensuring that all site staff under their direction perform their duties in compliance with the project quality assurance system. They are the main link between head office, the client and the project, and as such it is their responsibility to ensure the successful completion of the project to the complete satisfaction of the client.

The site manager ensures that all work carried out conforms to specified standards and design criteria. They are charged with enforcing the quality assurance system throughout the project, with due regard to the performance of subcontractors, ensuring conformity to specified standards by all subcontractors and suppliers.

Any discrepancies in information or flaws in design should be notified to the project administrator for their consideration as part of the quality assurance process. All internal and external correspondence, on-site checks and matters arising which might affect the successful completion of the project should be fully documented and acted upon in accordance with the procedures manual.

Any action necessitated by the quality assurance documentation process must be initiated and progress monitored in the appropriate manner. All quality procedures initiated have to be communicated to the quality manager for inclusion in the company's quality assurance system.

Site engineer

The site engineer has responsibility for setting out of the works as directed by the site manager, and they must carry out their duties in full compliance with stated specifications and design criteria and to fully document all their site activities in the required manner, keeping detailed records of

- site levels;
- verbal directions from the client's engineer;
- discrepancies in information supplied by the client's representative; and
- events necessitating a deviation from current designs or specifications, reporting any non-conformity directly to the site manager.

Quantity surveyor

The main responsibility of the quantity surveyor is the financial control of the project, but during the performance of their measurements for payments they should note any defective workmanship or non-conformance with the required standards or specifications. They should keep the site manager informed of any such items, discussing possible causes and any remedial action required.

Trade and contract foremen

All trades foremen and subcontractors' foremen have a responsibility to ensure that work carried out by persons under their direction is to the required standard as specified by the site manager, with due regard to plans, schedules and specifications, reporting any non-conformance directly to the site manager. All defects or non-conformances identified during the course of their work must be documented in accordance with procedures for submission to the site manager.

Stores foreman

The stores foreman's responsibilities include the checking of all invoices and materials for correctness to specifications when delivered to the site. They should also ensure the correct storage of all materials, reporting defects or non-conformances arising from deviations in procedures directly to the site manager. These should also be fully documented.

Site operatives

Site operatives have a responsibility to perform their works to the required standard and as specified by their foreman or site manager. During the execution of their duties, they should report defective materials or contradictions of information to their foreman or the site manager.

Generic outline of a project quality assurance system

The project quality assurance file details the project particulars for use in the effective management of the project. The quality assurance system needs to be documented in a way which makes all project information easily accessible to the user and ensures consistent use of the quality assurance documentation procedures for all site functions. This can be achieved by clear and methodical indexing of all quality procedures documentation.

Project quality assurance file overview

Project details

This section on project details lists the general information required for reference purposes by the site management team. This will consist of the following information:

- details on the client
- project address

- brief description of the project
- commencement date
- completion date.

Contract directory

The contract directory lists all persons or organisations who are involved in the project, detailing their involvement and giving their addresses, and telephone and fax numbers. This information, together with the project details, is essential for the effective communication of the management team and the co-ordination of all on-site participants/activities.

Site management structure

This is usually presented as an organisational tree or flow chart indicating all site staff, showing the lines of authority and autonomy, giving reference to any individual or department at head office which occupies a supporting role to individual site staff.

Subcontractors

This section details all subcontractors on site, providing details of foreman, anticipated labour force and any pre-contract undertakings or agreements, for example specialist lifting equipment or machinery, to be supplied by the main contractor. Details of head office organisation and individuals directly responsible for the subcontractors are also noted.

A subcontractor's programme is included and requires incorporating into the main contractor's programme.

Materials specifications

The materials specifications section details all material types to be used on the project, noting all details of required specifications, storage and handling details, control of substances hazardous to health (COSHH) assessments, and any additional considerations, for example lengthy order times or details of especially expensive and fragile items.

Project programme

A detailed project programme will be included, giving full particulars of start and finish dates of all contractors and subcontractors, durations, available float times, minimum and maximum resources, and projected valuations.

Quality policy

The quality policy details the company's objectives and commitment to quality and states the standard to which the company's quality system conforms, for example BS/EN/ISO 9001. It is imperative that all staff fully understand this document and that it is implemented, as it will form the basis of the third party audit conducted by the certification body.

Construction Design and Management (CDM) regulations

The company's CDM file must be represented in the project quality assurance file and will include details of the following:

- method statements
- plant and equipment identification
- hazard risk assessment
- subcontractors' risk assessment and method statements
- COSHH and risk assessment records
- Health and Safety Executive (HSE) notification of project
- company health and safety (H&S) policy
- company insurers
- details of emergency procedures.

Full documentation of the above will not be contained within the project quality assurance file, but a brief summary of each entry, with specific reference to the full documentation of each, is necessary.

Quality assurance documentation control files

The quality assurance documentation files contain all the necessary documents for the implementation of the quality assurance system, ensuring that all site activities are documented and quality procedures adhered to. All documents contained within the quality assurance documentation files are relevant to the administration of the project and in maintaining quality procedures. As most construction projects are individual in design and specification, the quality assurance system needs to take account of the project's individuality. This is achieved by incorporating design criteria and specifications for the project into the quality assurance system, using this information as a form of work instruction and quality requirement, with checks performed on workmanship and materials to ensure conformance. A typical quality assurance documentation file cover sheet is shown in Figure 4.10.

Details of quality assurance documentation

A typical cover sheet for this section is provided in Figure 4.11.

Document issue record

The document issue record details all project documents issued to the main contractor from the tender stage onwards, providing document numbers, revision details, dates and origin of all documents. The issue records all documentation issued by the main contractor to subcontractors or other parties. A copy of the document issue record is sent to head office each week as part of the company's quality assurance system.

A pro forma suitable for recording document issues is provided in Figure 4.12. Further pro forma examples will succeed each document referred to in this section and be identified by appropriate figure numbers.

Best Construction

PROJECT
QUALITY ASSURANCE FILE

CONTENTS
PROJECT DETAILS
CONTRACT DIRECTORY
SITE MANAGEMENT STRUCTURE
SUBCONTRACTORS
MATERIAL SPECIFICATIONS
PROJECT PROGRAMME
QUALITY POLICY
C D M

Fig. 4.10 Project quality assurance documentation file front cover: example pro forma

Best Construction

**QUANLITY ASSURANCE
DOCUMENTATION FILE**

CONTENTS
DOCUMENT ISSUE RECORD
COMMUNICATIONS
INFORMATION RECEIVED
ARCHITECT'S INSTRUCTION REQUESTED
ARCHITECT'S INSTRUCTIONS RECEIVED
VERBAL INSTRUCTIONS

Fig. 4.11 Quality assurance documentation details cover sheet: example pro forma

Best Construction

BEST CONSTRUCTION					
DOCUMENT ISSUE RECORD					
Sheet Number		Best Construction			
Client					
		74 Main Street			
Contract Number		Main Town			
Contract Title		DN11100			
& Address		Tel. No.			
		Fax No.			
Document Type and Description	**Date of Issue**	**Date of Amendments/Revisions**			
Distribution		**Number of Copies**			

Fig. 4.12 Document issue: example pro forma

Drawing issue record

The drawing issue record is used in the same instances as the document issue record detailed above, and a pro forma is shown in Figure 4.13.

Communications

All communications received must be photocopied and date-stamped, and details of origin and actions required must be noted with each communication.

Best Construction

BEST CONSTRUCTION				
DRAWING ISSUE RECORD				
Sheet Number		Best Construction		
Client				
		74 Main Street		
Contract No.		Main Town		
Contract Title		DN11100		
& Address		Tel. No.		
		Fax No.		
DWG No.	**DWG Title and Description**	**Date of Issue**	**Date of Amendments/Revisions**	
Distribution		**Number of Copies**		

Fig. 4.13 Drawing issue: example pro forma

All communications sent must be photocopied and recorded in the same manner. Photocopies of all correspondence should be forwarded to head office, with a file copy retained on site. Actions required need to be undertaken immediately and documented accordingly. A typical pro forma is shown in Figure 4.14.

Information requested

Requests for information must be made in writing using the appropriate form and providing all necessary information, including the project details, the person from whom the information is being requested, the nature of the information, the date of despatch and the signature of the person receiving the request.

Best Construction

BEST CONSTRUCTION		
COMMUNICATION RECEIPT AND DISPATCH		
Client	Best Construction	
Contract No.		
Contract Title	74 Main Street	
Contract Address:	Main Town	
	DN11100	
	Tel. No.	
	Fax No.	
Details	**Received**	**Dispatched**
Signature of Site Manager		**Date**

Fig. 4.14
Communications
receipt and despatch:
example pro forma

Two copies are retained, one on site and one at head office. Figure 4.15 provides an example pro forma.

Information received

All information received needs recording, with details of the date received, the origin of information, and the name and position of the person issuing the information. If the information is in response to a request for information, details of the request form must be given for reference purposes. On receipt of information, a copy is sent to head office and a copy retained on site. A typical pro forma is indicated in Figure 4.16.

Architect's instruction requests

Alterations to the design and/or specification of the project must be preceded by an architect's instruction (AI). Requests for AIs need to be made in writing

Best Construction

BEST CONSTRUCTION	
Information Request Form **Ref. ....**	
Client	Best Construction
Contract No.	
Contract Title	74 Main Street
	Main Town
Contract Address	DN11100
	Tel. No.
	Fax No.
Signature of Site Manager	**Date** ....
Details of Information Requested	
Signature and Position of Person in Receipt of Request	

Fig. 4.15 Information request: example pro forma

using the appropriate form, noting the date, the nature of the request and project details, and signed by the recipient of the request form. A copy of the AI request form is sent to head office and a copy retained on site; an example can be found in Figure 4.17.

Architect's instructions received

On receipt of an AI, it must be referenced to any AI request form previously issued and a copy forwarded to head office, with a copy retained on site. The course of action necessitated by the issue of an AI will depend upon its content; for example, the AI may request extra work or may omit an item of work, therefore any further documentation will be at the discretion of the site manager. See Figure 4.18 for an example.

Best Construction

BEST CONSTRUCTION	
Receipt of Information Form	
Ref.	
Client	Best Construction
Contract No.	
Contract Title	74 Main Street
	Main Town
Contract Address	DN11100
	Tel. No.
	Fax No.
Signature of Site Manager	**Date**
Details of Information Required	

Fig. 4.16 Receipt of information: example pro forma

Verbal instructions

Where time constraints necessitate a verbal instruction, this has to be fully documented stating full details of the instruction, the date of issue, and the name and position of the person providing the instruction, see Figure 4.19 for an example. Any verbal instruction has to be followed up by a request for an AI, which is then documented as previously described.

Subcontractor audit

Subcontractor audits are conducted at approximately monthly intervals to monitor conformance with the quality assurance system, company policy and the subcontractor's own quality system, work standards and conformance to

Best Construction

BEST CONSTRUCTION	
Request for Architect's Instructions **Ref. ....**	
Client	Best Construction
Contract No.	
Contract Title	74 Main Street
	Main Town
Contract Address	DN11100
	Tel. No.
	Fax No.
Signature of Site Manager	**Date** ....
Details of Information Required	
Signature and Position of Person in Receipt of Request	

Fig. 4.17 Request for architect's instructions: example pro forma

any pre-tender requirements. A standard pro forma should be employed, and an example is shown in Figure 4.20.

Supplier audit

The supplier audit is usually performed at monthly intervals and monitors supplier conformance with the company's quality assurance system and the supplier's own quality system, materials specifications and level of service. The results of the audit are forwarded to head office, with a copy retained on site. Any non-conformance is acted upon by the quality manager. Figure 4.21 provides an example.

Best Construction

BEST CONSTRUCTION	
Receipt of Architect's Instructions	
Ref.	
Client	Best Construction
Contract No.	
Contract Title	74 Main Street
	Main Town
Contract Address	DN11100
	Tel. No.
	Fax No.
Signature of Site Manager	**Date**
....	....
Details of Information Required	

Fig. 4.18 Receipt of architect's instructions: example pro forma

Materials invoice check

Materials invoice checks are conducted by the stores person upon delivery of all materials. Figure 4.22 provides an example pro forma. Reference is made to the materials specification documents in the quality assurance documentation file, which are then cross-checked against the delivery invoices to verify the correct specification of materials. All data pertaining to the delivery are logged in the materials delivery register, with any non-conformance fully documented. The completed register is submitted to the site manager for quantitative checks. A copy is sent to head office and one retained.

Best Construction

BEST CONSTRUCTION	
Verbal Instruction	
Ref.	
Client	Best Construction
Contract No.	
Contract Title	74 Main Street
	Main Town
Contract Address	DN11100
	Tel. No.
	Fax No.
Signature of Site Manager	**Date**
....	....
Details of Information Required	
Signature and Position of Person Issuing Instruction	
....	

Fig. 4.19 Architect's verbal instructions: example pro forma

Correspondence

All correspondence must be recorded, and the most efficient method is to photocopy all correspondence and detail the date of receipt or despatch, the origin or destination and the name and position of the person sending or receiving the correspondence, and provide a reference to any further actions necessitated by the correspondence, e.g. requests for AIs.

Weekly reports

Weekly reports are written by all site management staff and submitted to the site manager. These weekly reports refer to the working standards of

Best Construction		
Interim Audit: Subcontractors		
Name of Subcontractor		
Title of Contract		
Contract Number		
Project/Contracts Manager		
1. Have subcontracts fully complied with our pre-tender requirements?	YES	NO
2. Has the standard of workmanship required been maintained?	YES	NO
3. Has a punctual and regular attendance according to programme been maintained?	YES	NO
4. Have the subcontractors co-operated with other trades?	YES	NO
5. Have all contract details and deadlines been conscientiously pursued?	YES	NO
Comments By		

Fig. 4.20 Internal audit for subcontractors: example pro forma

subcontractors and employed labour. The quality of materials, accuracy of project drawings and design information, and details of any events requiring a deviation from the design criteria are documented. Figure 4.23 shows a typical pro forma.

Minutes of meetings

Copies of all minutes are kept on site for reference purposes, and any actions required by the minutes are documented in the appropriate manner.

Other documentation

Other documentation for use in the efficient and effective management of the project will include general administrative documents used for day-to-day management of the company and project. These include, for example:

- time sheets
- plant records

Best Construction

Best Construction		
Interim Audit: Suppliers		
Name of Suppliers		
Title of Contract		
Contract Number		
Project/Contracts Manager		
1. Have suppliers fully complied with our pre-tender requirements?	YES	NO
2. Have all deliveries been to the required specifications in terms of		
	Timeliness?	
	Quality?	
	Quantity?	
	Service level?	
3. Has the materials supplier been co-operative with our and subcontracting organisation(s)?		
Comments By:		

Fig. 4.21 Interim audit of suppliers: example pro forma

- Subcontractor attendances
- Weather condition records
- Scaffold register
- Accident report record.

Commentary The implementation of a carefully designed quality assurance system can enhance a company's performance and promote good working practices, which in turn will improve the future prospects of a company and, ultimately, the industry.

Total quality management

This chapter has thus far concentrated on BS/EN/ISO 9000 series quality systems and their application to a construction operational environment. However,

Best Construction

Best Construction					
Material Delivery Register Ref.					
Client			Best Construction		
Contract Number			74 Main Street		
Contract Title			Main Town		
Contract Address			DN11100		
			Tel. No.		
			Fax No.		
Signature of Stores Person			**Date**		
Material/ Product	**Quantity Delivered**	**Quantity Accepted**	**Quantity Rejected**	**Supplier**	**Reason for Rejection**
Comments By					

Fig. 4.22 Materials delivery: example pro forma

some construction organisations have applied total quality management (TQM). TQM is a philosophy for achieving a never-ending improvement through people. TQM has had limited success in construction firms, and the reasons for failure have a high positive correlation with the reasons for failure associated with BS/EN/ISO 9000 series quality systems. In order to provide a comprehensive chapter on quality assurance systems applied to construction organisations, this section on TQM has been included. The following establishes the main problems associated with the implementation of TQM in a construction operational environment. A generic implementation model designed to

Best Construction

Best Construction	
Weekly Report Ref.	
Client	Best Construction
Contract No.	74 Main Street
Contract Title	Main Town
Contract Address	DN11100
	Tel. No. ....
	Fax No. ....

Position.............................	Signature	Date..................................
Item/Issue	Action	Comments

Fig. 4.23 Weekly report: example pro forma

assist construction enterprises in the attainment of a successful TQM process is incorporated.

Rationale for the implementation of TQM in a construction operational environment

The fundamental rationale for implementing TQM in a construction operational environment is the attainment of a sustainable competitive advantage. TQM has been advocated as a strategy for achieving an improvement in the effectiveness, flexibility and competitiveness of construction-related enterprises. Oakland and Aldridge (1995) identified that the construction industry is associated with a patchy reputation for the quality of its products and services, with most projects not being completed on time. TQM aims to produce a superior performance from the whole project team. This results in improved quality of products and services, delivery and administration, which ultimately satisfies the client's functional and aesthetic requirements within defined cost and completion parameters. Ghobadian and Gallear (1996) conducted research which established that the performance of companies that had implemented TQM exceeded their industry's median performance. However, the implementation process can be a most problematic activity encompassing many pitfalls for an unwary organisation. The following section establishes the problems associated with the implementation of TQM in a construction operational environment. Advocated solutions to the problems of implementation are offered for consideration.

Problems of implementation

Two out of three organisations engaged in the TQM implementation process consider it a failure. This high failure rate is due to the following factors; although listed separately they are not mutually exclusive, and most organisations experience a combination.

Insufficient commitment by senior management

Senior management must instil in all employees of the host organisation a desire to improve the competitiveness of the company. TQM's three vital elements are systems, people and resources. Successful implementation is dependent upon senior management developing and organising these key elements. Oakland (1993) advocates that TQM 'requires total commitment, which must be extended to all employees at all levels and in all departments'. Therefore, senior management must be fully committed to the implementation processes, as can be evidenced by senior management providing all the resources required for the TQM initiative.

Inappropriate corporate culture

TQM requires a corporate culture based on trust and a desire to identify problems in order to eliminate them, thus improving production processes. The concept of 'empowerment' is a vital part of the TQM philosophy. If a climate of mistrust exists between senior management and the rest of the organisation, the implementation process is doomed to fail. Organisations must understand that a truly 'morphogenic' change is necessary and that a 'morphostatic' change will not sustain TQM. Organisational culture dictates the way a business operates and how employees respond and are treated. Organisational culture contains such elements as a guiding philosophy, core values, purpose and operational beliefs. These elements have to be integrated within a mission statement which translates the cultural theory into tangible targets bounded by closed objectives.

No formal implementation strategy

The implementation process should be planned. TQM is a project and therefore requires planning as a project. Treating it as a bolt-on organisational activity will lead to failure. TQM is a means of improving the competitiveness, effectiveness and flexibility of an entire organisation. Achieving these noted advantages requires organisations to plan and organise every operational activity at all levels. This process must be part of strategic implementational development and not treated in isolation. Senior management must also understand that the benefits of implementation are not instantaneous: TQM is a long-term corporate investment.

Lack of effective communications

The life blood of any organisation is communication, and the importance of this organisational activity cannot be over-emphasised. Within a TQM framework, all employees of the host company should be able to communicate as required. Do not forget the concept of 'internal' and 'external' customers, with its requirement for effective communication mechanisms. If employees are to become part of the organisation's decision-making process, they need a means of expressing their views to senior management. Control in any organisation is dependent upon the communication systems function.

Narrowly based training

The key to successful TQM implementation is having staff who are competent to execute their allocated tasks. If employees are empowered to plan and perform work activities, it is vital that they possess all the necessary skills and competencies required. A primary function for a construction-related enterprise seeking to gain a competitive advantage is to implement 'training and education in teamwork' (Hellard, 1993). As an example, if staff are to participate in group discussions, training in group dynamics and public speaking would be advantageous.

Concentrate on organisational strengths

TQM is designed to provide a competitive advantage based upon the host organisation's strengths. Senior management should not lose sight of the fact that sustained competitive advantages are obtained by implementing strategies that exploit their strengths by responding to environmental opportunities while neutralising external threats and avoiding internal weaknesses. According to Barney (1991), the following two standard corporate planning techniques can be used: a strengths, weakness, opportunities and threats (SWOT) analysis; or a politico-legal, economic, socio-cultural and technological (PEST) analysis.

Key elements for consideration that lead to successful implementation of TQM

- Senior management must obtain a full understanding of the philosophy and requirements of TQM, because they are responsible for establishing a quality-focused organisation.
- A common vision is required by all employees of the host organisation. This can be accomplished by adopting awareness sessions, customer surveys, benchmarking and common vision workshops.
- Provision of the necessary resources, which include human as well as financial requirements, and education and training for quality improvements.

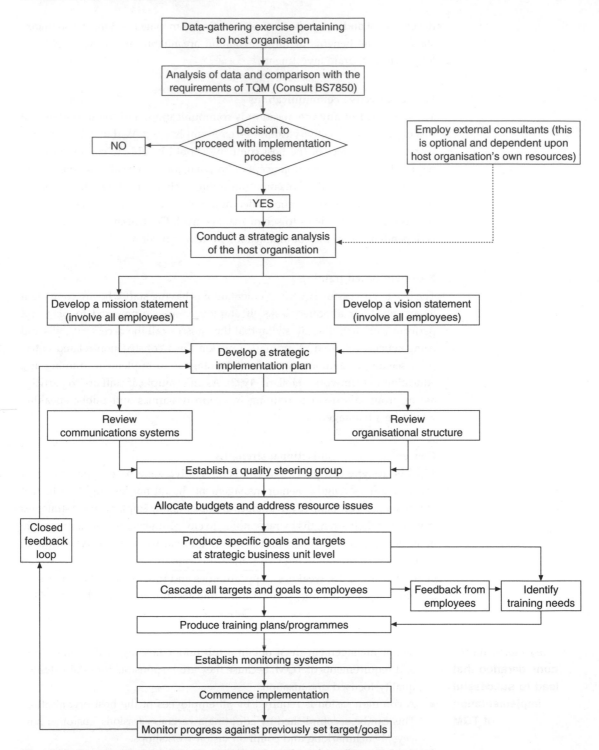

Fig. 4.24 Generic model for the implementation of total quality management

- The development of an implementation strategy which may be based on an incremental process. Senior management must review quality management systems in order to maintain progress.
- Designing procedural systems pertaining to work practices. Concentration of organisational effort should be on preventive rather than corrective actions.

Summary

The UK government has acknowledged the problems of the construction industry by accusing it of lacking customer focus and being ready to use any excuse to pursue so-called claims against government (*The Guardian*, 8/11/95). TQM enables construction companies to fully identify the extent of their operational activities and focus on customer satisfaction. Part of this service focus is the provision of a significant reduction in costs through the elimination of poor quality in the overall construction process. If a construction organisation can overcome the implementation problems, then a sustained competitive advantage is the prize. Figure 4.24 provides an overview of the TQM implementational process and should be of great value to construction organisations pursuing TQM.

It should be noted that the BSEN ISO 9000 series of standards is being amended. The new set is to be published in November 2000, therefore the content of the new standards will not be known until after publication. However, the broad aims of the new standard are the same as the current BSEN ISO 9000 series.

References

Barney, J. (1991) 'Firm resources and sustained competitive advantage', *Journal of Management*, 17 (1), 99–120.

British Standards Institution (1972) BS 4891: *A Guide to Quality Assurance*, London.

British Standards Institution (1978) *A National Strategy for Quality*, Milton Keynes.

British Standards Institution (1987) *British Standard Number 5750*, British Standards Institution, London.

British Standards Institution (1989) *The Way to Capture New Markets*, British Standards Institution, London.

Construction Industry Research Association (1989) *Quality Management in Construction. Certification of Product, Quality and Management Systems*, SP 72, DTI, London.

Dalton, J. B. (1988) *Quest for quality: developments in the management of quality by the United Kingdom*, Department of the Environment, Property Services Agency, Vol. 1, CIB W/65, Organisation & Management of Construction Proceedings.

Ghobadian, A. and Gallear, D. N. (1996) 'Total quality management in SMEs, *Omega International Journal of Management Sciences*, 24 (1), 83–106.

Glassman, A. M. (1978) *The Challenge of Management*, J. Wiley & Sons, Toronto.

Griffith, A. (1990) *Quality Assurance in Building*, Macmillan Education, London.

The Guardian (1995) 'Millions are wasted as controls on prestigious projects fail', 8 November p. 4.

Hellard, R. B. (1993) *Total Quality in Construction Projects*, Thomas Telford, London.

Liebmann, J. D. (1993) 'A quality initiative postponed', *New Directions For Institutional Research*, 78, pp. 117–21.

Lozier, G. and Teeter, D. (1993) 'Six foundations of total quality', *New Directions For Institutional Research*, 78, pp. 5–11.

Majaro, S. (1992) *Managing Ideas for Profit*, McGraw-Hill, Maidenhead.

Morris, D. and Brandon, J. (1993) *Re-engineering your Business*, McGraw-Hill, London.

Oakland, J. S. (1990) *Enterprise Initiative Statistical Process Control*, Department of Trade and Industry, London.

Oakland, J. S. (1993) *Total Quality Management*, Butterworth-Heinemann, London.

Oakland, J. S. and Aldridge, A. J. (1995) 'Quality management in civil and structural engineering consulting', *International Journal of Quality & Reliability Management*, 12 (3), 32–48.

Passmore, W. A. (1994) *Creating Strategic Change*, J. Wiley & Sons, London.

Peters, T. (1988) *Thriving on Chaos*, Macmillan, London.

Pick-Up (1990) *Quality Management for Further Education*, East Midlands Further Education Council.

Property Services Agency (PSA) (1986) *Quality Assurance*, Department of the Environment.

Roche, J. D. (1980) *National Survey of Quality Control in Manufacturing Industry*, University of Galway, pp. 1–52.

Sayles, L. R. (1989) *Leadership Managing in Real Organisations*, McGraw-Hill, London.

Shewart, W. A. (1931) *Economic Control of Quality of Manufactured Product*, Macmillan, New York.

Vander Erve, M. (1993) *The Power of Tomorrow's Management*, Butterworth-Heinemann, London.

Weber, M. (1908 [1968]) *Economy and Society*, translated and edited by G. Roth and C. Witrich, Irving Publications, New York.

5 Health and safety regulation and implementation systems

Introduction

> The construction industry has a poor health and safety record. Serious injury and death happen regularly as a result of construction work. This affects not only workers, but also members of the public. Good management of the construction process is essential to help prevent accidents and ill health in the industry.
>
> (HSE, 1994a)

The construction industry is, by its very nature, hazardous. It has experienced an incident rate, in respect of fatal accidents and serious injuries, virtually unparalleled by any other industry. *Blackspot Construction*, the definitive report by the Health & Safety Executive (HSE), places the safety record of the industry in a clear perspective. In the period 1981 to 1985 there were over 130 fatal accidents each year on construction sites. Moreover, several thousand accidents resulting in serious and minor injury were recorded each year, although it is well accepted that many accidents which occur within the construction industry go unreported.

Health and Safety Commission (HSC) statistics for the 1990s indicate a year-by-year reduction in fatal accidents in construction. However, in the period 1990 to 1996 the average number of fatalities each year still exceeded 80 (see Figure 5.1). There can be no doubt, therefore, that construction remains a hazardous occupation.

The government has recognised and the industry has accepted that the accident record of construction might be improved considerably where responsibility for safety is assumed throughout all stages of a construction project by the various contractual parties. The Construction (Design and Management) Regulations 1994, which became effective from 31 March 1995, legislate to meet this aspiration. Commonly referred to, in abbreviated form, as the CDM Regulations, they apply to construction projects and everyone involved with them: clients, consultants, contractors, subcontractors, site operatives, building owners and occupiers. 'They do not apply to every project or everyone all of the time, but most projects and people are affected' (HSE, 1994a). Put simply, the CDM Regulations are concerned with the management of health and safety throughout the construction process.

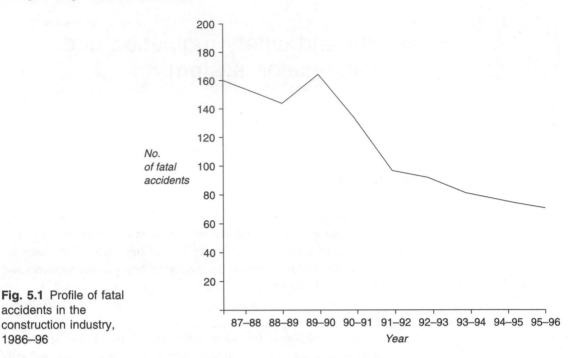

Fig. 5.1 Profile of fatal accidents in the construction industry, 1986–96

The Regulations place responsibility on clients, designers and contractors to be proactive in the planning, co-ordination and management of health and safety. Working together as a team, each participant must undertake specified key duties to ensure that a safe working environment is delivered and maintained.

The CDM Regulations are not intended to be prescriptive, and they do not specify performance standards. In considering the management of health and safety for any construction project, they focus on identifying the potential hazards to health and safety through each stage of the process, together with the assessment of their risk.

This chapter provides an insight into the concepts and implementation of the CDM Regulations. In synopsis, aspects explained are the legislative framework; roles and key duties of project participants and compliance with CDM during the design, planning and management of a construction project. The concepts, principles, management systems and procedures described are illustrated by a complete practice example and safety management system development pro forma for practice reference.

BS 8800: specification for health and safety management systems

The focus of this chapter is to outline the fundamental requirements for establishing a health and safety management system which meets the practical requirements of the CDM Regulations. In terms of management systems for health and safety, certificated in the same way as quality and environmental management, one is concerned with BS 8800, the UK's specification for health and safety management systems (H&SMS).

Recently introduced, certification bodies cannot, at the time of writing, provide BS 8800 accredited certification services for H&SMS. However, it is clear that certification bodies are working rapidly towards providing accredited certification, which will see H&SMS recognised in the same way as quality and environmental management systems.

Health and safety law

The legal system   Health and safety law, of which the Construction (Design and Management) Regulations 1994 are but one aspect, apportions duties and responsibilities to all parties involved in construction. Health and safety law is founded in both statute law – law made by Acts of Parliament – and Common Law, determined by judicial precedent.

An appreciation of the legal system within which the law is applied and administered is a prerequisite to understanding how health and safety regulation is developed and implemented. The law can be categorised according to how it is applied and administered into two main systems – criminal and civil. The essential difference between these systems is in the legal processes that are used and the resulting outcomes. The intended outcome of criminal proceedings is punishment, whereas the outcome of civil proceedings is remedy, for example compensation. Criminal cases are normally brought before the courts by an official of the state, the HSE being one such authority, while a civil case may be brought by any individual or organisation which, as a consequence of a breach of the law, has suffered injury, damage or loss.

Criminal and civil liabilities can arise from the same event. This may occur in construction where, for example, an employee seeks compensation for injury from the employer, while the employer is brought before the courts for a breach of the law. Whereas an employer can insure against civil liability, with the insurer paying compensation to the injured employee, one cannot insure against criminal liability, where punishment is the object of the outcome.

The legal systems are based on sources of law – rules used to bring cases to court and determine their outcomes. These are now described.

Common law

Common law is founded upon the recorded decisions of judges on cases which are brought before the courts. The principle of *judicial precedent* means that judgments in previous legal cases guide the case being heard. Judicial precedent applies to both criminal and civil proceedings, and a decision taken in a higher court usually binds a decision made in a lower court.

Although statute law has legislative supremacy over common law, principles of common law coexist with statutory legislation. This link is significant in health and safety law, where a *duty of care* is implicit in any contract of employment. This common law principle exists alongside statutory legislation for health and safety matters, within which most resides. A duty of care is most significant to construction contracts where this principle influences, for example, the relationship between client and contractor, or employer and employee.

A duty of care means that one party must take reasonable care to ensure that something they do, or something they could do but decide not to, will not cause harm to the other party as a result of their action or inaction. *Reasonable care* can be interpreted as the care taken by a reasonable person who exercises reasonable consideration before doing something which may, through their action, harm another person. In law, failure to take reasonable care where a duty of care is owed can lead to a case of *negligence*.

Statute law

Statute law is legislation founded by Acts of Parliament. In some Acts, government ministers have powers to create detailed legislation, termed *enabling acts*, leading to *statutory instruments* or *regulations*. One of the most prominent examples of an enabling act is the Health and Safety at Work, etc. Act 1974, under which the CDM Regulations were founded. There are many Acts of parliament and Statutory Instruments which affect the construction industry. One of the first was the Factories Act 1961 and one of the most comprehensive, the aforementioned Health and Safety at Work, etc. Act 1974.

Acts of Parliament

The Factories Act 1961

The Factories Act 1961 gave rise to a number of regulations influencing the construction industry:

- The Construction (General Provisions) Regulations 1961.
- The Construction (Lifting Operations) Regulations 1961.
- The Construction (Working Places) Regulations 1966.
- The Construction (Health and Welfare) Regulations 1966.

Milestones in health and safety legislation, these regulations have now largely been subsumed within and been superseded by more recent regulations.

The Health and Safety at Work, etc. Act 1974

The Health and Safety at Work, etc. Act 1974, commonly referred to as the HSWA, established the Health and Safety Commission (HSC), which is responsible for proposing policy and regulations, and the Health and Safety Executive (HSE), which has the responsibility for enforcing health and safety legislation. The HSWA is the principal legislative Act of Parliament under which almost all health and safety regulations, including the CDM Regulations, have been made.

European Directives and UK legislation

EU Directives

On 1 January 1973, the United Kingdom became a member of the European Community (EC) as a result of the European Communities Act 1972. Following the Maastricht Treaty on European Union 1993, the EC became the European Union, or EU. As a member of the EU, the UK must adhere to all EU legislation, with some of the most stringent exercised over health and safety. EU legislation becomes law in the UK through Acts of Parliament and enabling acts, giving rise to statutory instruments and regulations. These were outlined earlier.

EU legislation takes four forms:

1. *Directives* – these establish minimum standards for legislation imposed by EU member countries. The requirements set by Directives are mandatory, although the mode of implementation is left to the discretion of the national government.
2. *Regulations* – these apply directly to member countries and automatically become a part of national legislation.
3. *Decisions* – these are binding deliberations of the European Court of Justice on the subject in question.
4. *Recommendations* – these convey the opinions of bodies or institutions on the subject in question and as such are only advisory.

Within EU health and safety legislation, Article 118A of the Treaty of Rome (1957), introduced by the Single European Act (1986), requires that member countries 'pay particular attention to encouraging improvements, especially in the working environment, as regards the health and safety of workers'.

Born from this requirement, Directive 89/391/EEC, commonly referred to as a 'Framework Directive', so-called because it provides a framework for other Directives, was established. This Directive has been enacted in the UK by the Management of Health and Safety at Work Regulations 1992.

UK Legislation

The Management of Health and Safety at Work Regulations 1992 (MHSWR), often referred to within construction as the 'six-pack', enacts a set of Directives implementing health and safety law (see Figure 5.2). The Directives are as follows:

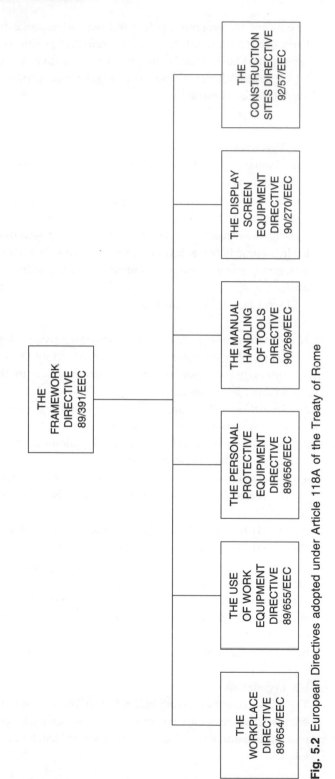

Fig. 5.2 European Directives adopted under Article 118A of the Treaty of Rome

- *Directive 89/654/EEC*: the 'workplace Directive' – minimum health and safety requirements for the workplace.
- *Directive 89/655/EEC*: the 'use of work equipment Directive' – minimum health and safety requirements for the use of equipment at work.
- *Directive 89/656/EEC*: the 'personal protective equipment Directive' – minimum health and safety requirements for use by workers of personal protective equipment in the workplace.
- *Directive 90/269/EEC*: the 'manual handling of loads Directive' – minimum health and safety requirements for the manual handling of loads where there is a risk of injury.
- *Directive 90/270/EEC*: the 'display screen equipment Directive' – minimum health and safety requirements for work with display screen equipment.

These Directives have been implemented by the following statutory instruments (see Figure 5.3).

- The Workplace (Health, Safety and Welfare) Regulations 1992 (SI 1992/3004), otherwise known as 'the Workplace Regulations'.
- The Provision and Use of Work Equipment Regulations 1992 (SI 1992/2932), otherwise known as 'the Work Equipment Regulations'.
- The Personal Protective Equipment at Work Regulations 1992 (SI 1992/2966), otherwise known as 'the Personal Protective Equipment (PPE) Regulations'.
- The Manual Handling Operations Regulations 1992 (SI 1992/2793), otherwise known as 'the Manual Handling Regulations'.
- The Health and Safety (Display Screen Equipment) Regulations 1992 (SI 1992/2792), otherwise known as 'the VDU Regulations'.

The sixth Directive within the Framework Directive, Directive 92/57/EEC: the 'Construction Sites Directive', which sets minimum health and safety requirements for temporary or mobile construction sites, has also been adopted under Article 118A of the Treaty of Rome. This is implemented in the UK by the Construction (Design and Management) Regulations 1994, otherwise known as 'the CDM Regulations', making up the six elements of the six-pack regulations.

The MHSWR require that employers undertake the following:

- carry out a risk assessment to determine the risks to health and safety of their employees and others.
- make arrangements to implement preventive or protective measures to mitigate the risks identified.
- monitor activities for the occurrence of risks.
- appoint competent persons to *manage* risk.
- provide health and safety information to employees (and to temporary and visiting workers).

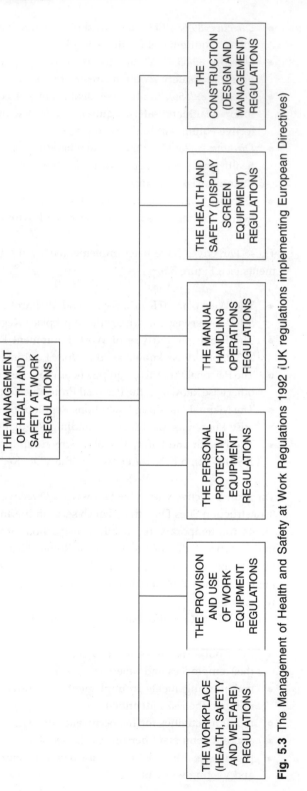

Fig. 5.3 The Management of Health and Safety at Work Regulations 1992 (UK regulations implementing European Directives)

It is the prominent requirement for conducting *risk assessment* that lays the foundation for health and safety implementation systems in construction. While there is no specified requirement to carry out risk assessment in the CDM Regulations, it is a duty of all parties involved with the construction processes. Moreover, general responsibilities under other legislation, for example the MHSWR, demand that risk must be considered.

The Construction (Design and Management) Regulations 1994

A European study of construction industry fatal accidents showed that although the primary cause of 37% of the accidents was failures of the construction site management and workers, 28% of accidents could be attributed to poor planning and 35% due to unsafe design. An important conclusion was therefore that over 60% of accidents were due to decisions made before the work began.

(Croner, 1994)

Recognising the significance of this study, Directive 92/57/EEC: on the implementation of minimum safety and health requirements at temporary or mobile construction sites, was implemented. This Directive placed responsibility on everyone involved with construction projects to address the associated risks. In the UK, the Directive was enacted through the introduction of the Construction (Design and Management) Regulations 1994 (the CDM Regulations).

The Regulations specifically and clearly emphasise the *management* of health and safety in the construction processes of planning, design and production. Whereas in the past the contractor assumed responsibility for health and safety, the CDM Regulations make all parties contribute to health and safety management.

Application of the Regulations

The CDM Regulations apply to:

1. any 'project' that involves 'construction work' that is 'notifiable' (these terms are described subsequently).
2. any project that is not notifiable but where five or more persons will be involved at any one time.

Project

A project means 'a project which includes or is intended to include construction work'.

Construction work
Construction work is (Croner, 1994):

> the carrying out of any building, civil engineering or engineering construction work and includes any of the following:
>
> (a) the construction, alteration, conversion, fitting out, commissioning, renovation, repair, upkeep, redecoration or other maintenance, decommissioning, demolition or dismantling of a structure. (structure defined subsequently)
>
> (b) the preparation for an intended structure, including site clearance, exploration, investigation (but not site survey) and excavation, and laying or installing the foundations of a structure.
>
> (c) the assembly . . . or . . . disassembly of a pre-fabricated structure.
>
> (d) the removal of a structure or waste resulting from demolition or dismantling of a structure.
>
> (e) the installation, commissioning, maintenance, repair, or removal of mechanical, electrical, gas, compressed air, hydraulic, telecommunications, computer or similar services which are normally fixed within or to a structure.

Structure
Structure means (Croner, 1994):

> (a) any building, steel or reinforced concrete structure . . . railway . . . or . . . tramway line, dock, harbour, inland navigation (i.e. canal) tunnel, shaft, bridge, viaduct, waterworks, reservoir . . . pipeline . . . , cable, aqueduct, sewer, sewage works, gasholder, road, airfield, sea defence works, river works, drainage works, earthworks, lagoon, dam, wall, caisson, mast, tower, pylon, underground tank, earth retaining structure or structure designed to preserve or alter any natural feature.
>
> (b) any formwork, falsework, scaffold or other temporary structure designed or used to provide support or means of access during construction work.
>
> (c) any fixed plant from which a person can fall more than two metres during installation, commissioning, de-commissioning, or dismantling work.

Notifiable
Notifiable projects are those where:

- the production stage on site will exceed 30 days; or
- the production stage on site will exceed 500 person-days.

Where projects are notifiable, the planning supervisor must write to the HSE and provide the following information:

- date of notification
- construction project address
- name and address of client(s)* or client's agent*
- type of project, i.e. construction work type
- name and address of the planning supervisor*
- signed declaration by the planning supervisor of appointment
- name and address of the principal contractor*
- signed declaration by the principal contractor of appointment
- commencement date for start on site
- planned duration of the works
- estimated maximum number of persons to be at work on the site
- planned number of contractors to be at work on the site
- name and address of contractors already appointed.

definitions of the various parties and their duties under the Regulations follow subsequently.

Exemptions to the application of the Regulations

There are a number of situations where the Regulations do not apply. These include works undertaken by a local authority or domestic householder, and where the project is not notifiable by the criteria previously specified. Figure 5.4 illustrates the application of and exemptions from the Regulations.

Key features of management

The CDM Regulations place a duty upon clients, consultants and contractors to coherently and methodically think about their contributions to project health and safety. Moreover, their inputs must be co-ordinated and managed throughout all stages of the construction process. This involves consideration for health and safety in project evaluation and development, design (design engineering), production, maintenance and repair, and de-commissioning (and or demolition).

The key features of health and safety management within the CDM Regulations are:

- *Risk assessment* – parties must identify and assess project health and safety risk to comply with their duties.
- *Competence and adequate resources* – every designer, contractor and planning supervisor (described subsequently) must be pre-qualified by assessment to ensure that they are competent and have the necessary resources to fulfil their duties for health and safety.
- *Co-operation and co-ordination* – all parties have an obligation to co-operate and co-ordinate their efforts to identify and mitigate health and safety risks.
- *Provision of information* – all parties have a duty to share information pertinent to health and safety for development of the project's 'Health and Safety Plan' and 'Health and Safety File' (described subsequently)

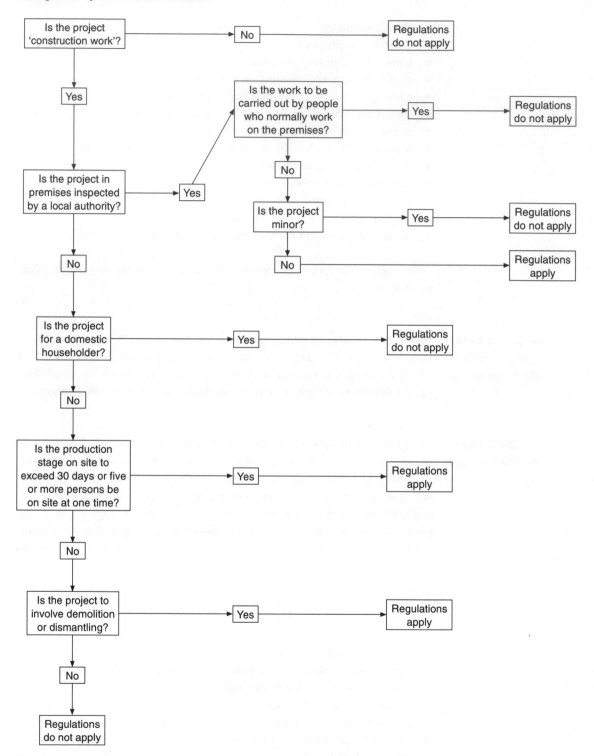

Fig. 5.4 Application of the Construction (Design and Management) Regulations 1994

Roles and duties of project participants

The planning authority and building control

It is sometimes asked if there is any link between the implementation of the CDM Regulations and obtaining planning permission for a construction project. The CDM Regulations neither assign responsibility nor place any duty upon planning approval or building regulatory authorities for health and safety matters.

Planning approval and building control are explicitly concerned with giving due consideration to the nature, type, location and siting of buildings and structures which are allowed to be constructed. Such authorities are not concerned with the methods of construction and, therefore, their remit does not encapsulate health and safety. Responsibilities for project health and safety rest clearly with the participants: the client, consultants and contractors, together with those responsible for ensuring that health and safety laws are met, i.e. the Health and Safety Executive (HSE).

The Health and Safety Executive

The HSE has the responsibility for enforcing the CDM Regulations. The HSE is also responsible for enforcing the duties and responsibilities of clients, designers and planning supervisors. It is therefore likely that any or all of these positions will be visited during the course of a contract, generally to ensure that duties are being undertaken or specifically to investigate an accident.

HSE visits to construction sites are a well-established procedure, and this continues under the CDM Regulations. Inspectors may request to see a construction health and safety and plan and also inspect the site to ensure that the plan is appropriately implemented.

The client

The client may be defined as 'any person for whom a project is carried out.' The client may appoint an agent to act on their behalf, and to all intents and purposes under the Regulations the agent is treated as the client. An agent must be assessed as competent and have the necessary resources to carry out the duties of the client they represent. When an agent is appointed, the arrangement must be notified in writing to the HSE.

The client's key duties are:

- to determine if the project must comply with the CDM Regulations;
- to appoint a *planning supervisor* and a *principal contractor* for the project (these parties are described subsequently) and ensure that both are competent and have adequate resources to fulfil their obligations;
- to ensure that any design consultants and contractors that they nominate are competent and have adequate resources;
- to provide the planning supervisor with information pertinent to health and safety matters on the project;

- to ensure that the construction work on site does not commence until an appropriate health and safety plan has been prepared; and
- to ensure that the project's health and safety file is available for perusal following completion of the project.

The planning supervisor

The planning supervisor may be defined as 'a person appointed by the client who has overall responsibility for co-ordinating the health and safety aspects of the design and planning phase'. The planning supervisor, rather than being an individual, can be a multi-disciplinary team where the project is large and complex, for example.

The client may fulfil the role with an in-house organisation, or the principal contractor may be appointed in the capacity. Nevertheless, it is the client's responsibility to ensure that the appointee is competent and has adequate resources.

There must be a planning supervisor in place throughout the duration of the project, although the appointee can be changed at the discretion of the client as the project proceeds. The appointed person(s) and any changes made must be notified in writing to the HSE.

The planning supervisor's key duties are:

- to ensure that designers comply with their duties, in particular the identification, avoidance and mitigation of risk and also in providing adequate information on health and safety for construction, maintenance and demolition;
- to encourage co-operation between the designer and other designers and/ or other consultants for the purposes of promoting health and safety;
- to ensure that a health and safety plan is prepared before the principal contractor is appointed;
- to give advice, if requested, to the client on the appointment of designers and contractors, and to advise contractors appointing designers;
- to advise the client on the health and safety plan before construction work commences;
- to notify the HSE of health and safety management arrangements for the project; and
- to ensure that the health and safety file is complete and passed to the client at the end of the project.

The designer

The designer may be defined as 'the person(s) who prepares designs (or engineers) or arranges for persons under their control to develop a design'. The designer should ensure that the health and safety of those persons involved in construction work, maintenance and repair are considered during the formulation of design to reduce aspects of risk on the project.

The designer's key duties are:

- to ensure that the client is aware of his/her duties under the Regulations;
- to develop designs which avoid risks to health and safety;
- to consider, when formulating designs, risks which may arise during construction or maintenance;
- to reduce the level of risk where it is not possible to eliminate the risk;
- to consider measures which will protect workers where avoidance or reduction of risk to a safe level is not possible;
- to ensure that the design includes adequate information on health and safety and to pass this information on to the planning supervisor for inclusion in the health and safety plan; and
- to co-operate with the planning supervisor and, where necessary, other designers to enable them to comply with health and safety legislation.

The principal contractor

The principal contractor may be defined as 'a person who undertakes or manages construction work or who arranges for persons under their control to undertake or manage construction work'. The client has a duty to appoint a principal contractor for the project as soon as is practicable. The client can be the principal contractor providing they are, by definition, a contractor.

The principal contractor has to take over and develop the health and safety plan and co-ordinate the activities of all contractors such that they comply with health and safety legislation. The principal contractor's key duties are:

- to develop and implement the health and safety plan;
- to arrange for competent and adequately resourced contractors to undertake work where it is subcontracted;
- to ensure the co-ordination and co-operation of all contractors;
- to determine from contractors the findings of their risk assessments and details of how they propose to carry out high-risk operations;
- to provide information to contractors of project site risks;
- to ensure that contractors and operatives comply with site rules and procedures specified in the health and safety plan;
- to monitor health and safety performance throughout the project;
- to ensure that all operatives are fully informed and consulted on health and safety matters;
- to allow only authorised persons on to the site;
- to display in a prominent position the notification of the project to the HSE; and
- to ensure that information is passed to the planning supervisor for inclusion in the health and safety file.

Contractors

A contractor may be defined as 'a person who undertakes or manages construction work or who arranges for persons under their control to undertake or manage construction work as subcontractors of the principal contractor'. Contractors include those persons who are self-employed. Contractors are required to co-operate with the principal contractor and provide information concerning health and safety risk assessment and management procedures.

The contractor's key duties are:

- to provide information for the health and safety plan concerning the risks arising from their work, together with details of the actions they will take to manage those risks;
- to undertake their work while complying with directions from the principal contractor and with rules and procedures specified in the health and safety plan;
- to provide information for the health and safety file and report to the HSE under the Reporting of Injuries, Diseases and Dangerous Occurrences Regulations 1995 of instances of dangerous occurrences, ill health and injuries; and
- to provide adequate information to their employees on health and safety matters.

Where the self-employed are themselves employers they assume the aforementioned key duties. All employers must provide employees with the name of the planning supervisor, the name of the principal contractor, and details of the health and safety plan.

Employees

An employee may be defined as 'a person employed by and under the direct control of an employer'. It is the responsibility of the principal contractor to check that all employees on the project have been given adequate health and safety information and training.

All employees are entitled to information concerning health and safety matters during the production process and should be given the opportunity to express their views to the principal contractor.

Relationship of key duties to health and safety law

The duties of the project's participants fall into three categories of responsibility under health and safety law:

1. *Absolute* – this is a duty that *must* be carried out. It imposes an absolute obligation on a party, and any breach of duty may result in prosecution.
2. *Practicable* – this is a duty that should be carried out irrespective of inconvenience, time or cost. The standard of performance is high but not absolute.

3. *Reasonably practicable* – this is a duty that is carried out having considered the balance of that duty against inconvenience and cost involved. Where cases of breach of duty are brought, it is the responsibility of the accused to demonstrate that it was not reasonably practicable to have done more than that done to comply with the duty.

Duties which must be carried out so far as reasonably practicable occur frequently in health and safety legislation. For example, within the CDM Regulations the designer's key duties, as presented previously, are to be carried out 'in so far that they are reasonably practicable'.

Appointment of the planning supervisor, consultants and contractors

The contractual arrangements between the client and the planning supervisor, design/engineering consultants and contractors set out the terms agreed between them for providing services to the project. The terms specify the duties and obligations of the parties and moreover, determine responsibilities. Before the introduction of the CDM Regulations, contracts and appointments rarely addressed in detail the roles and duties of the parties with respect to health and safety. The CDM Regulations, however, do bring the roles, duties and responsibilities of each party sharply into focus.

The CDM Regulations specify duties which must be fulfilled. The parties have an obligation to determine how and by whom these duties will be undertaken, and this must be clearly understood between them. Where a project comes within the remit of the CDM Regulations, the specified duties for health and safety are imposed irrespective of any agreed obligations within the contractual arrangement between the parties. If the CDM Regulations do not apply to the project, while there would be no liability under CDM for any breach of health and safety law, there may still be liability if a case were to be proved under the Management of Health and Safety at Work Regulations 1992.

Appointment of the planning supervisor

The CDM Regulations require that the planning supervisor be appointed by the client as soon as is practicable and at an early stage to help the client to fulfil his duties. The client should monitor and review progress of the planning supervisor's duties, as the appointment can be terminated and a new appointment made should there be any lapse in competence or failure to resource the task adequately.

It was mentioned previously that the planning supervisory role must be maintained throughout the project, but there may be situations where a change in planning supervisor is deliberately scheduled. In design–build procurement for example, the planning supervisor may be appointed to deliver services during the conceptual design stage and a different planning supervisor appointed from the design–build principal contractor at the detailed design stage.

It was also mentioned previously that the CDM Regulations do not impose 'absolute' responsibility but rather responsibilities that are 'practicable' or 'reasonably practicable'. If absolute responsibilities are required by a client then they should be specified unambiguously in the contract arrangement.

In the CDM Regulations, the planning supervisor is required to ensure that a pre-tender health and safety plan is compiled. It does not state a duty upon the planning supervisor to prepare the plan himself. If the client wishes the planning supervisor to prepare the plan then the client would need to specify this. Similarly, the planning supervisor has to ensure that the health and safety file is compiled, but there is no responsibility unless specifically determined for the planning supervisor to compile it. There is, therefore, a clear need for the client and planning supervisors to consider carefully what duties are to be performed both generally under the Regulations and in meeting any additional requirements expressed by the client.

The Royal Institute of British Architects (RIBA) can provide the client with a 'Form of Appointments as Planning Supervisor' for use when arranging the services of construction professionals as planning supervisors. The form consists of a memorandum of agreement between the client and planning supervisor setting out a schedule of services to be provided by the planning supervisor.

Appointment of consultants

Design consultants have especially onerous responsibilities under the CDM Regulations. Whereas other contractual parties are influenced by exemptions to the application of CDM, the designer's duties apply irrespective of the size and type of project or the number of staff employed.

The RIBA has a 'Standard Form of Agreement for the Appointment of an Architect', which incorporates a supplement to include the application of the CDM Regulations. The document sets out conditions of appointment and services expected to meet health and safety requirements.

A further RIBA document, 'Conditions of Engagement for the Appointment of an Architect', presents guidance on the responsibilities of the architect where acting only as the designer, i.e. where a planning supervisor is appointed and the designer does not act in that capacity.

Appointment of contractors

The appointment of the principal contractor is a significant element in ensuring that health and safety is effectively managed during the construction phase. The consideration of health and safety in the selection and appointment process is a prerequisite to the development of management systems and procedures which follow during the works on site.

The client should ensure that any prospective principal contractor is competent, has considered health and safety in their tender and included for health and safety management in their tender bid price. A client may take measures to ensure this, first by adopting a health and safety pre-qualification assessment of all prospective principal contractors, and second by inviting a detailed health and safety policy to be submitted for consideration.

There are no set procedures which must be followed for health and safety during the tender stage. However, 'A Guide to Managing Health and Safety in Construction' (HSE, 1995b) is available as a source of advice to clients.

From the client's perspective, several key points should be noted:

- The tender documents which are sent to prospective principal contractors should be formatted in such a way that the methods by which the principal contractor includes for and prices health and safety management is transparent.
- The tender documents should include a pre-tender health and safety plan that will inform the prospective principal contractor of significant risks which have been identified and which should be considered in the tender in terms of management requirements and cost implications.
- The tendering period should be sufficient to allow prospective principal contractors to consider the health and safety management aspects comprehensively and tender appropriately.

From the prospective principal contractor's perspective, they should be able to demonstrate that they:

- are competent to manage health and safety on the project. This can be achieved by providing records from previous projects undertaken and profiles of qualified staff to be employed on the project.
- have adequate resources to manage the health and safety matters which are identified. This can be demonstrated in the costings provided in the tender price.
- have implemented a health and safety policy within their organisation which is applied to their projects. This should be available in existing company documentation such as statements of annual reports.
- have formal procedures for managing those health and safety aspects identified in the pre-tender health and safety plan.

In considering and evaluating those tenders submitted, the client may take advice from the planning supervisor, as outlined in the planning supervisor's key duties. In addition, clients and planning supervisors may meet and discuss in further detail a prospective principal contractor's services to the project.

Project health and safety documentation

Two significant documents evolve from the management of health and safety under the CDM Regulations:

1. The health and safety plan – developed in two parts:
   - the pre-tender health and safety plan; and
   - the construction phase health and safety plan.
2. The Health and Safety File.

The health and safety plan

The health and safety plan provides the health and safety focus for the construction phase of a project. The pre-tender health and safety plan should be prepared in time so that it is available for contractors tendering or making similar arrangements to carry out or manage construction work. The planning supervisor is responsible for seeing that this is done. After being appointed by the client, the principal contractor needs to develop the health and safety plan and keep it up to date.

(HSE, 1994a)

The pre-tender health and safety plan

The pre-tender health and safety plan should include:

- a general description of the project and works;
- the time-scales for completion;
- details of known health and safety risks to operatives;
- information required by contractors for them to demonstrate that they are competent and have adequate resources;
- the information needed by the principal contractor to develop the construction-phase health and safety plan;
- information required by contractor on compliance with welfare provisions.

The construction-phase health and safety plan

The construction-phase health and safety plan should include:

- the arrangements for ensuring health and safety of all persons who may be affected by the project;
- the arrangements for the management of health and safety of the project and monitoring of compliance with health and safety law;
- information concerning welfare arrangements for the project.

The health and safety file

This is a record of information for the client/end user, which tells those who might be responsible for the structure in future of the risks that have to be managed during maintenance, repair or renovation.

(HSE, 1994a)

The planning supervisor must ensure that the file is prepared as the project progresses and is handed to the client when the project is complete. The client must ensure that the file is available to persons who work on any future design, construction, maintenance and repair, or demolition of the structure.

Management of health and safety

The CDM Regulations are directly concerned with the planning for, encouragement for, teamwork in and co-ordination of the management of project health and safety. The CDM Regulations recognise the complexities characteristic of the construction processes. In addition, they acknowledge the traditional separation of the participants that come together to form the project team, together with the challenges and problems that this can bring.

The essential aim of the CDM Regulations is therefore to ensure that health and safety are consciously considered by all project participants. Furthermore, to ensure that health and safety become intrinsic to the management of the construction project from the outset, the CDM Regulations demand compliance with a two-phased approach to health and safety planning, the development of (1) the pre-tender health and safety plan; and (2) the construction health and safety plan. It is this requirement which forms the basis for a systems management approach within which *risk assessment* is the central theme.

The main parties to the project employed by the client, namely the design consultant and principal contractor, each play an essential part in developing health and safety management systems and working procedures. These identify, assess and control risk both within and across their professional boundaries. In addition, consistent with good systems management, there are information feedback loops both within the span of control of the parties' activities and across the construction processes. In the context of the CDM Regulations, this ensures a full contribution by each party to health and safety planning, the development of implementation systems, and project review, as evidenced through the delivery of the health and safety file (see Figure 5.5).

Risk assessment *Risk assessment* is a well-established and recognised analytical technique used widely across many different fields of business, commerce and industry. In the context of health and safety in construction, its application focuses upon the identification, assessment and control of risk to minimise those hazards which can arise in the course of undertaking a project.

The CDM Regulations emphasise risk assessment within the duties and responsibilities of both the designer and the principal contractor. For example, a designer is required to avoid foreseeable risks to health and safety of any person carrying out construction work; to combat risks at source; and to give priority to measures which protect the whole workforce.

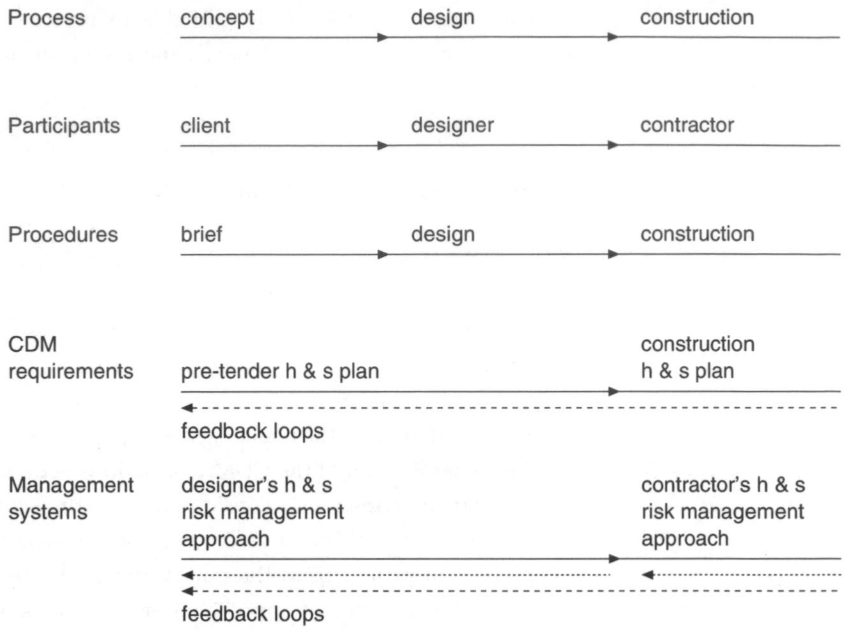

Process	concept	design →	construction
Participants	client	designer →	contractor
Procedures	brief	design →	construction
CDM requirements	pre-tender h & s plan →		construction h & s plan

← -

feedback loops

| Management systems | designer's h & s risk management approach | | contractor's h & s risk management approach |

← · ← · · · · · · · · · · · · ·
← -

feedback loops

Fig. 5.5 Management systems for project health and safety to accommodate CDM Regulations

Risk assessment for construction health and safety involves three key activities:

1. hazard identification
2. evaluation of risk
3. prevention and protection measures.

Hazard identification

A *hazard* is 'something which presents a potential to cause harm'. This could be through the occurrence of an accident or exposure to a dangerous situation, material or substance. In the construction industry, ever-present hazards leading to fatal and serious injury commonly involve working at heights; use of ladders and scaffolds; collapse of temporary structures; use of vehicles, mechanical plant and equipment; and exposure to harmful substances. Hazard identification involves the systematic recognition of any aspects of a project which have a potential to be a danger to those persons working on or being around that project.

Evaluation of risk

Risk is 'the likelihood that a specified undesired event will occur due to the realisation of a hazard' (Croner, 1994). Once a hazard has been identified, the *degree of risk* must be determined. Two factors are influential in this determination: (1) *the severity of harm* – the level of harm that a circumstance would create; and (2) *the likelihood of occurrence* – the frequency of a hazardous circumstance.

Table 5.1 Evaluation criteria for severity of harm

Assigned value	Description
1	minor injury – no first aid attention
2	illness – chronic injury
3	accident – needing first aid attention
4	reportable injury – under RIDDOR*
5	major injury – under RIDDOR*
6	death

* RIDDOR: Reporting of Injuries, Diseases and Dangerous Occurrences
Regulations 1995

Table 5.2 Evaluation criteria for likelihood of occurrence

Assigned value	Description
1	remote – almost certain not to occur
2	unlikely – occurrence in exceptional circumstances
3	possible – certain circumstances would influence occurrence
4	likely – could ordinarily occur
5	probably – high chance of occurrence
6	certain – 100% chance of occurrence

A risk assessment gives the statistical probability of a hazardous event occurring. The outcome is based on a body of information, qualitative and quantitative, from factual experience to develop a numerical figure, which represents the degree of risk. The following example illustrates this.

Suppose a situation indicates that the *severity of harm* is 3 and the *likelihood of occurrence* is 4, where the severity of harm is an assigned value on a six-point scale from minor injury (1) to death (6), based on factual information (see Table 5.1), and the likelihood of occurrence is an assigned value on a six-point scale from remote (1) to highly probable (6), based on factual information (see Table 5.2).

degree of risk = severity of harm × likelihood of occurrence

$$= 3 \times 4 = 12$$

The *degree of risk* (12) is a numerical value which is a proportion of the possible maximum degree of risk. The maximum risk is 36 (the severity of harm on the six-point scale, multiplied by the likelihood of occurrence on the six-point scale). The value (12) can, perhaps, be more meaningfully expressed as a percentage of the maximum risk (36), i.e. 33%. This is the percentage chance of the hazardous event occurring.

Table 5.3 Determination of priority rating for risk

Severity of harm	Likelihood of occurrence		
	HIGH / LOW 10% MEDIUM / LOW 5% LOW / LOW 1%	HIGH / MEDIUM 50% MEDIUM / MEDIUM 25% LOW / MEDIUM 5%	HIGH / HIGH 100% MEDIUM / HIGH 50% LOW / HIGH 10%

In risk assessment, rather than expressing the degree of risk in percentage terms, a *priority rating* is given, for example:

- low priority (L)
- medium priority (M)
- high priority (H)

where, in the example, valued criteria give the following priority bands:

- L = 3–9%
- M = 10–44%
- H = 45–100%

However, bands on the points scale can vary according to the value assigned to the criteria. Therefore, in the example, the risk of 33% would lie in the medium priority (M) band.

An alternative approach is to grade risk on a scale of 1 to 10, where 1 = 1% and 10 = 100%. A table can be developed to determine the degree of risk where the severity of harm and likelihood of occurrence have been assigned priority ratings: low, medium and high (see Table 5.3).

The evaluation of risk enables the value to be calculated for any hazard identified. The greater the value, the higher the priority and therefore the more thought that should be given to, and effort that should be placed upon, avoiding or managing the risk.

Prevention and protection measures

When an evaluation of risk has been considered, the principles of prevention and protection should be applied . . . The principles, in summary are to: (a) avoid risk; (b) combat risk at source; (c) control risk.

(Croner, 1994)

For example, a requirement of the CDM Regulations is for the designer to identify risks associated with the construction work and to redesign to avoid those risks occurring.

Where avoiding risk proves impossible, the designer must make provision to combat the risk and ensure that appropriate information is passed to the planning supervisor and thereby to the principal contractor. This will allow safe systems of work to be developed which will control the level of risk from those hazards identified.

Pre-tender health and safety plan

The CDM Regulations state that the management of health and safety for any project commences with planning, design and specification. Clients and designers can, therefore, make very considerable contributions in identifying hazards and assessing and managing risks by formulating a clear and comprehensive health and safety plan early in the project development sequence.

The pre-tender health and safety plan is a key element in the identification and assessment of risk. The information gathered is essential to project evaluation and development. Furthermore, the information is a prerequisite to developing the construction health and safety plan, which lays the foundation for a safe working environment on site.

The pre-tender health and safety plan should contain information such that prospective principal contractors tendering for the project can plan and cost for safety measures in their price. Where a contract is awarded or negotiated rather than bid for, the same information should still be provided for this first stage of health and safety planning.

Contents of the pre-tender health and safety plan

The information required for developing a pre-tender health and safety plan will be gathered within three broad aspects of the project:

- the existing environment
- the design
- the site.

Information will be input by the client, designer and planning supervisor (see Figure 5.6). The information is project-specific and therefore will vary from one project to the next. Nevertheless, information can be gathered under standard headings to simplify matters.

'Managing Construction for Health and Safety', the HSE-approved code of practice, suggests nine headings under which information for the pre-tender health and safety plan might be gathered:

1. nature of the project
2. existing environment
3. existing drawings
4. design
5. materials

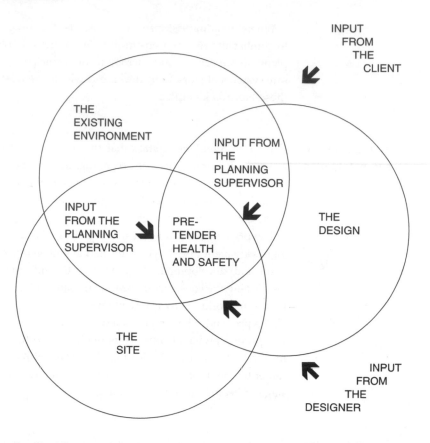

Fig. 5.6 Inputs to the development of the pre-tender health and safety plan

6. site elements
7. overlap with the client's activities
8. site rules
9. continuing liaison.

Nature of the project

This section will contain such information as the name and address of the client, the location of the project site, details of the construction, and timeframe for the construction phase.

Existing environment

This section will describe the environment on and around the project site. This is important to the consideration of site access and egress and to activities taking place at the interface between the site and its environs, for example materials delivery, or the movement of heavy mechanical plant. Details of existing service utilities, for example water, electricity and telephone, together with possible concealed services such as gas pipes or water mains, should be included as these may represent a potential safety risk. Existing buildings and structures should also be included in the plan as these could pose a possible hazard.

Existing drawings

Existing drawings may be available from landowners, building owners or occupiers, utility companies, or local authorities. These may record site layouts, structural details of existing or neighbouring properties and location of existing services. Any such drawings and specifications should be included in the plan. If the project involves work to recently completed existing buildings or structures, there will be available health and safety files from which useful information might be drawn.

Design

Findings of the designer's risk assessment should be included in the plan. Any hazardous works should be identified and described and requests for method statements highlighted for the attention of the principal contractor.

Materials

Details of any potentially hazardous materials specified in the design and for which specific precautions will be required should be given. This relates to particular materials noted as hazardous by the designer, for example special paints or adhesives. Materials which are generally hazardous, for example cement or plaster, should be implicitly understood as such by any competent construction operative.

Site elements

This section will include information which is important to any persons present on the site, for example temporary accommodation; delivery, unloading and storage; pedestrian and vehicular traffic routes; services and amenities; and site boundaries.

Overlap with the client's activities

Where the project involves work to existing premises there may be restrictions to working practices. For example, working hours may be limited, noise and pollution controls may be required beyond those usually encountered, or access may be limited to specific areas or times. Any such details should be included in the plan.

Site rules

The client, designer or planning supervisor may impose rules to be observed on site, for example vehicular speed restrictions or a no-smoking policy. These should be clearly stated in the plan.

Continuing liaison

The plan should contain details of key contacts for the client, designer and planning supervisor. Information will include the full name, address and telephone number of each contact. In addition, the protocol for requesting particular information, for example instructions, should be described.

Inputs to the pre-tender health and safety plan

The client

The client has an essential contribution to make to the pre-tender health and safety plan. This takes the form of providing information which is key to the project's development. Such information might include:

- location and description of the works
- time-frame for project development and construction
- existing documentation and drawings of the site
- details of site topography and conditions
- conditions of existing buildings and structures
- existence and location of temporary services
- arrangements for site access and traffic management
- need for security arrangements
- facilities for storage and protection
- potential restrictions to work routines and general operations
- findings of risk assessments.

These aspects represent some of the information that may be provided by the client and the consultants that they employ. Some information may be available from existing sources, while some might have to be acquired through undertaking specific surveys. The salient point is that all the information that is provided by the client is project site-specific. Details relating to the construction form are incorporated in the information provided by the designer.

The Designer

The designer will be able to provide information to the planning supervisor considering both the site and the project. Such information may include the following:

- details of existing buildings industry internal layouts
- results from soil investigation and analysis
- phasing and sequencing of the works
- hazards presented by the proposed design
- specific requirements for method statement preparation, for example hazardous elements of demolition operations or specification for temporary works.

Such information is essential to the potential principal contractor, who must consider, incorporate and resource for any significant risks within the pre-tender health and safety plan.

Construction-phase health and safety plan

Following the appointment of the principal contractor, the principal contractor assumes responsibility for the development of the construction-phase health and safety plan. On some projects, there will be project development and design works still being undertaken, although the principal contractor has been appointed. In this situation, these works remain the responsibility of the planning supervisor until they are at the appropriate stage to be passed to the principal contractor for inclusion in the construction-phase health and safety plan.

The information included in the pre-tender health and safety plan forms the basis of development for the construction-phase health and safety plan. As stated previously, the client has a responsibility to ensure that the health and safety plan is prepared subject to any outstanding aspects under the control of the planning supervisor, before construction commences on site.

In compiling the construction-phase health and safety plan, the principal contractor will provide information to enable:

- development of a framework for managing health and safety of all those involved in the construction stage of the project;
- the contributions of other organisations involved in the construction phase to be included;
- the organisation and action needed to investigate hazards, including informing all personnel who might be placed at risk;
- good communication by the contractual parties;
- the development of method statements, design details and specification for all project contractors (subcontractors);
- the security of the site to ensure that only authorised personnel gain access.

Contents of the construction-phase health and safety plan

'A Guide to Managing Health and Safety in Construction' (HSE 1995b) suggests a number of specific aspects which should be considered in compiling the construction phase health and safety plan. These are:

- *Project overview* – this section develops further the information contained in the sub-section of the pre-tender health and safety plan entitled 'nature of the project'.
- *Health and safety standards* – any standards specified by the client, the designer or principal contractor should be included.
- *Management arrangements* – this section should outline the management structure and organisation for the health and safety management of the project and specify the key responsibilities of the parties.
- *Contractor information* – this outlines the protocol involved in advising contractors about project risks.
- *Selection procedures* – procedures for assessing the competence and resources of contractors are included in this section.

- *Communication and co-operation* – this section sets out lines of communication and methods for co-ordinating health and safety.
- *Activities with a risk to health and safety* – hazards identified in the pre-tender health and safety plan are to be communicated to site personnel. This section details how information is to be explained and how hazards can be mitigated or reduced.
- *Emergency procedures* – this section outlines notification of alarms, escape routes, assembly areas and personnel checks.
- *Accident recording* – this section outlines how the contractor will fulfil his responsibilities under RIDDOR (HSE, 1995a).
- *Welfare facilities* – the arrangements for all temporary site welfare facilities are included in this section.
- *Training* – this section outlines all health and safety training, including induction training.
- *Site rules* – any site rules and restrictions identified in the pre-tender health and safety plan should be involved in the construction-phase health and safety plan.
- *Consultation* – this section details the procedures for personnel to raise health and safety issues with their supervisors.
- *H & S file* – this details the procedures for the passing of information from the principal contractor to the planning supervisor.
- *H & S monitoring* – procedures for inspection and audit are outlined in this section.
- *Project H & S review* – a report on health and safety and details of any incidents should be outlined in this section.

Inputs to the construction-phase health and safety plan

To facilitate the detailed development of the construction-phase health and safety plan, the principal contractor will require inputs from other project participants. These include the following.

Subcontractors

Subcontractors can contribute to the construction health and safety plan by providing information concerning:

- identified hazards occurring from their services;
- the assessment of risks identified; and
- the potential measure of control of the identified risks.

Subcontractors are duty-bound to make their personnel aware of all project risks and provide training for those employees where necessary.

Designers

Some elements of the design will be finalised only after the construction stage on site has begun. Within the range of the CDM Regulations, the designer continues to have responsibilities for the design irrespective of when the design work is undertaken. Therefore, although the principal contractor assumes responsibility for the construction health and safety plan, and may be implementing it on site, the designer retains responsibility for any design elements still to be passed on to the principal contractor.

Such elements may include, for example, foundations, where there have been variations to those works originally proposed; services, which may have to be changed as a result of unforeseen circumstances; or finalisation of lift installation as a result of detail changes to fabrication design.

The client

The client will contribute to the development and possibly the implementation of the construction-phase health and safety plan. The client will therefore be included in discussion concerning production matters throughout the construction stage. For example, the principal contractor may request from the client further information on site rules or restrictions, access provision, or storage of potentially hazardous materials or substances. Also, the client may be asked to assist with site health and safety training of personnel. Nevertheless, it is the responsibility of the principal contractor to ensure that the construction-phase health and safety plan is fully developed and implemented throughout the project.

The designer's health and safety management system

The focus on risk assessment
In construction, the design process progresses through a sequence of development stages, which are:

1. project feasibility and outline concept
2. scheme design and layout
3. detail design and specification.

The most effective time to consider project health and safety and eliminate potential hazards is early in the design process, during project feasibility and outline conceptual development. As a construction project progresses, the opportunity to design out the hazards will diminish. Risks identified during conceptual development will allow the designer time and space in the design development schedule to determine mitigation measures, while risks identified during the detail design stage are likely to have no such latitude. The best that might be achieved at a later stage is that control measures can be suggested.

It is imperative that the key technique of risk assessment, as described earlier, is utilised from the earliest stage of the design process. As the design evolves through the various stages of the design process so risk assessment is

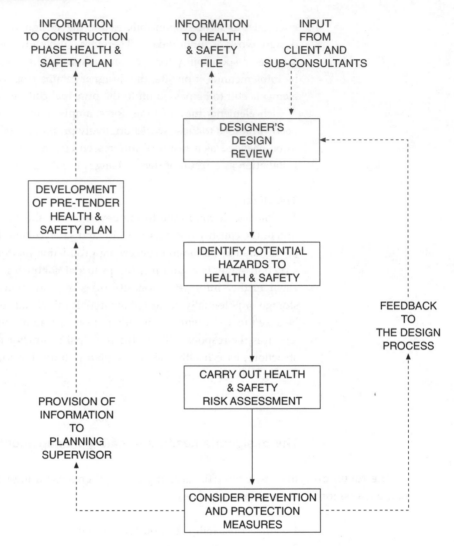

Fig. 5.7 Designer's health and safety management system during project evaluation and development (design review)

an evolving process. As a project's design develops it will be reviewed periodically to ensure that it meets the client's brief and the designer's intentions. Design review provides an excellent opportunity to evaluate the project's health and safety dimension. It forms the basis of the designer's health and safety management system.

Design review   Figure 5.7 illustrates the fundamental aspects of a designer's health and safety management system. The system is based upon developing the project design within a management information loop. The system commences with the receipt of the information necessary to develop the design from the client and any contributing subcontractor. It concludes with the information necessary to take forward the pre-tender health and safety plan to tendering and the formulation of the construction phase health and safety plan.

Key procedures

Health and safety design review involves four key procedures:

1. *Risk identification*: identify potential hazards to health and safety. Each hazard identified within the design should be given a reference number and be clearly described.
2. *Risk assessment*: carry out health and safety risk assessment. An assessment is made of the potential risks associated with each hazard identified (this is achieved using the risk assessment strategy and procedures described previously).
3. *Risk response*: consider prevention and protection measures. Where possible action should be taken immediately in response to the identified hazard. This might be achieved through a redesign of the hazardous element. However, some hazards may not be allocated at the time of identification, and the designer will need to revisit the issues as the design evolves and more information becomes available. Some hazards may be recognised and recorded but left unresolved for the contractor to address. This would be the case where, for example, the hazard was posed not by the design element but through the methods used to carry out the works.
4. *Risk communication*: Provision of information to the planning supervisor. The parties who need to be informed of the hazards should be determined. This will invariably be the planning supervisor, as it is the planning supervisor who utilises the information in the preparation of the pre-tender health and safety plan. However, sub-consultants may need to be informed of hazards as it affects their work and therefore there must always be a design feedback loop in the designer's reporting system.

System records

Systematic gathering and recording of information concerning project risk is essential to the designer's health and safety design review. The documentation necessary to establish a systematic approach to health and safety design review and to meet the requirements of the CDM regulations can be considered in two key areas:

1. the designer's hazard identification record;
2. the designer's risk assessment record.

The designer's hazard identification record
Figure 5.8 illustrates a pro forma which can be used to record information for the designer's hazard identification. The information recorded should include the following:

DESIGNER'S HAZARD INDENTIFICATION			
PROJECT:			
DESIGNER:			
STAGE OF WORK:		DATE:	
REF. NUMBER	DESCRIPTION OF HAZARD	ACTION REQUIRED	PERSONS TO INFORM

NOTES:
For example – drawing reference
                – reference to specifications

PERSONS TO INFORM – KEY:
PS – Planning supervisor
SE – Structural engineer
ME – Mechanical & electrical engineer
QS – Quantity surveyor
LA – Landscape architect
PC – Principal contractor
R – Review later

Fig. 5.8 Designer's hazard identification record: suggested pro forma

- *Project, design, stage of work and date* – this information identifies the record, relating it to the particular construction project. In addition, it records the date and stage of work, thereby relating the record to changes that take place as the project progresses.
- *Reference number* – as each hazard is identified by the designer when formulating the design, it should be given a reference number. This is to ensure that each hazard can be traced throughout the design process as the various design reviews are carried out.
- *Description of the hazard* – each hazard identified should be briefly described.
- *Action required* – details of the action taken at the design review to eliminate or reduce the hazard should be described. If the hazard is to be the subject of a later review or the matter is to be passed to the principal contractor for attention, then this should be noted.
- *Persons to inform* – information on hazards identified will invariably need to be passed to other participants in the project, for example the planning supervisor, sub-consultants and the principal contractor. Any persons to inform should be recorded in this section. To aid this, a legend specifying the initials of the various parties can be given at the foot of the pro forma.
- *Notes* – any other information, for example references to drawings and specifications, should be described in the notes section.

The designer's risk assessment record

Figure 5.9 presents a pro forma which can be applied to recording information for the designer's risk assessment record. This form records information which allows judgments to be made about the degree of risk and what actions can be taken to reduce the risk. The information recorded should include the following:

- *Project, designer, date, sheet number* – as with the project identification record, the risk assessment record should give basic information relating the record to the particular project, designer and date of compilation.
- *Reference number* – this is the number which was allocated to the hazard as recorded in the hazard identification record.
- *Element of work* – the element of the work, or activity, should be briefly described, for example; excavation for basement of main office building.
- *Potential hazard/risk* – aspects of the design posing the potential risk shall be described, for example working in close proximity to electricity cables.
- *Persons at risk* – as different controls may be required for particular persons, these persons should be listed on the record, for example the public, site operatives and visitors to the site.
- *Risk rating* – an estimate of the level of the risk should be given. The method of determining the degree of risk – the severity of harm and the likelihood of occurrence – was described in a earlier section of this chapter.

DESIGNER'S RISK ASSESSMENT									
DESIGNER:			PROJECT:			DATE:		SHEET	OF
REF. NUMBER	ELEMENT OF WORK	POTENTIAL HAZARD/RISK	PERSONS AT RISK	RISK RATING			ACTION AT DESIGN STAGE	ACTION	RISK CONTROL POSSIBILITIES
				L	S	R		BY	
								WHEN	
RISK RATINGS:		S – Severity (low/medium/high)		L – Likelihood (medium/high)				R – Risk (Severity × Likelihood)	

Fig. 5.9 Designer's risk assessment record: suggested pro forma

- *Action at the design stage* – actions taken to eliminate or resolve the degree of risk by altering the design or recommendations to the principal contractor for inclusion in the contractor's method statement should be described.
- *Action taken* – where any actions are taken, the persons responsible for initiating the actions should be noted and the date when taken.
- *Risk control possibilities* – the designer may identify risks which require a particular management course and for which specific action must be taken by others, for example a particular construction method and sequence. Such information should be given in this column for attention by the principal contractor.

The provision of information

The designer is duty-bound by the CDM Regulations to avoid foreseeable risks, combat risk at source, protect the entire workforce and communicate appropriately on any risk to the project as a result of the design solution.

It is essential that the designer has identified all the potential hazards in the design, together with those that may occur during construction as a result of the design. Hazards must not only be recognised but be highlighted to other contractual parties and be priced for by tendering prospective principal contractors, subcontractors and suppliers.

The information collected by the designer during design review, presented in the hazard identification record and risk assessment record, will be passed to the planning supervisor. It will be included in the pre-tender health and safety plan and ultimately form part of the project health and safety file. The timely and effective provision of information throughout the design process is essential in providing accurate information for the pre-tender health and safety plan. Good information is vital, because the information will be used in so many subsequent health and safety management stages – the construction health and safety plan, health and safety management on site and the generation of the health and safety file.

In addition, closing the system loop is imperative to the designer, as much can be gleaned about the design process within the project situation for later reference on other projects. As with all management systems, the designer's health and safety management system will envolve through experience, leading to benefits both for the design consultants and for future clients and project teams.

The principal contractor's health and safety management system

Safe systems of work   The health and safety management system established by the principal contractor is centred on creating *safe systems of work*. Safe systems of work are

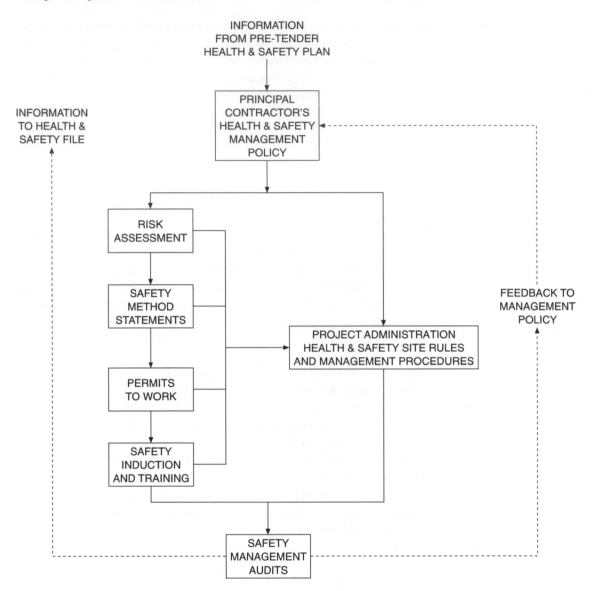

INFORMATION
FROM PRE-TENDER
HEALTH & SAFETY PLAN

INFORMATION
TO HEALTH &
SAFETY FILE

PRINCIPAL
CONTRACTOR'S
HEALTH & SAFETY
MANAGEMENT
POLICY

RISK
ASSESSMENT

SAFETY
METHOD
STATEMENTS

PROJECT ADMINISTRATION
HEALTH & SAFETY SITE RULES
AND MANAGEMENT PROCEDURES

FEEDBACK TO
MANAGEMENT
POLICY

PERMITS
TO WORK

SAFETY
INDUCTION
AND TRAINING

SAFETY
MANAGEMENT
AUDITS

Fig. 5.10 Principal contractor's health and safety management system

welfare- and safety-conscious procedures which are formalised by the implementing organisation, understood by all employees, practical in application and, above all, enforceable.

Figure 5.10 illustrates the basic components of a principal contractor's health and safety management system. The system consists of seven key elements. These encompass the principal contractor's duties and responsibilities for project health and safety under the HSWA 1974, the MHSWR 1992 and the CDM Regulations.

The system flows from incoming information from the pre-tender health and safety plan, through a series of managerial activities to auditing procedures.

Feedback mechanisms close the system loop with information directed to health and safety policy making. An external flow of information passes project information to the health and safety file, a requirement of the CDM Regulations described in an earlier section of this chapter.

Elements of the system

Seven key elements form the skeleton of the system. These are:

1. Health and safety management policies;
2. Risk assessment;
3. Safety method statements;
4. Permits to work;
5. Safety induction and training;
6. Project administration: site rules and management procedures;
7. Safety management audits.

Health and safety management policies

Management policies set out the organisation's statement of commitment to health and safety management. They outline how organisational procedures may be established to ensure that safe systems of work are developed and applied to the project. Policy formulation is fundamental to the development of any organisation's health and safety management system. It forms the basis for creating the managerial framework, the operational elements within the framework and the assignment of responsibilities to both managers and employees. To establish health and safety management, leadership and the motivation of employees, policies must be clearly stated and communicated. Policies must reflect the organisational commitment at corporate levels and pervade the whole organisation through the hierarchy of management to all site employees.

To achieve successful health and safety management at project level, the principal contractor must establish a clearly determined set of policies which specify basic procedures and expected minimum standards of performance where possible. Although particular characteristics of health and safety policies will differ among principal contractors and across construction projects, a generic outline can be suggested. A health and safety management system for any construction project should establish a manual of policies comprising each of the following aspects of the project:

- site access and registration procedures for employees and visitors;
- vehicular access to site and associated public safety;
- waste management and site egress;
- accommodation and welfare facilities;
- statutory notes and registers;
- hazard and safety signage;

- working at heights;
- personal protective equipment;
- occupational health (noise, vibration, hazardous substances, pollution, exposure limits to specific conditions, manual handling);
- fire prevention and incidents;
- use of electrical supplies, work around services and portable hand tools;
- appointment of safety supervisors;
- third-party and public safety;
- safety induction and training;
- safety method statements; and
- reporting injury, ill-health, near miss incidents and damage.

An example of stated policy for 'waste management and site egress' might be produced as follows:

Document Section A.1: Waste management
All contractors and personnel under their control shall be required to:
1. clear building debris and site waste which is the by-product of their operations as work proceeds.
2. maintain a designated and unobstructed route for egress of waste materials.
3. handle debris and waste to designated holding areas as specified in the construction-phase health and safety plan.
4. report any circumstances where there is any compromise of 1 and 2 above.

Risk assessment

It was seen earlier in this chapter that the designer is not required to devise safe systems of work for the principal contractor. This responsibility is held by the principal contractor. Risk assessment is a fundamental aspect of establishing safe systems of work and is a key element in the principal contractor's health and safety management system. The function of this system element is to provide management with details of potential hazards and risks and specify when safe systems of work are needed to control them.

Risk assessment concepts, principles and methodology were described in detail in a previous section of this chapter. Further guidance to risk assessment for safety management is given in the MHSWR 1992.

Safety method statements

Safety method statements provide details of how the safe systems of work are devised around construction operations and tasks. The statements inform the principal contractor's construction project management team of the proposed method of carrying out stage-by-stage tasks. For each task, the health and safety implications can be assessed and precautions taken to ensure a safe working environment. The function of this system element in the principal contractor's health and safety management system is to provide management with a formalised approach to linking the construction operational sequence to the health and safety dimension. This consideration is essential both to producing the principal contractor's construction phase health and safety plan and to assessing the risk of the proposed work of subcontractors. During the construction phase, safety method statements are an essential reference to guide on-going site activities.

A safety method statement enables operatives carrying out the work to:

- undertake their tasks in a safe manner;
- be aware of the hazards associated with their tasks; and
- implement control mechanisms to eliminate or reduce the risk of safety incidents.

To achieve these objectives, the safety method statement should:

- describe the operations and tasks that comprise the works;
- identify the location of the works;
- specify the supervisory arrangements while the works are on-going;
- identify the safety supervisor;
- list the plant needs for the operations;
- state likely occupational health implications of the work; and
- specify the precautions to be taken to minimise risk by those carrying out the works.

Permits to work

A permit to work is a management control procedure which can be issued for almost any construction activity. It is usually applied to activities that have a high risk of danger for those undertaking the work. The issue of a permit to work is a formalised procedure designed to provide a safe system of work, and as such it is issued by a section manager within the principal contractor's organisation. Essential aspects of a permit to work is that it always relates to tasks where:

- specific training is needed to undertake the task;
- the work is considered to be high-risk and denoted as such in the risk assessment records; and
- the works are complicated in nature or location.

This element of the health and safety management system concentrates on ensuring that the principal contractor has procedures in place to consider, issue, monitor and control permits for hazardous work.

The format of documentation procedures for issuing a permit to work will vary across organisations and even from task to task. Common aspects to be included are:

- a permit number
- date and time of issue
- the duration of the permit
- the location of the works
- a description of the hazard, likely risk or potential implications
- precautionary measures to be taken
- any testing or validating procedures of measures imposed
- emergency procedures, signals and reporting
- acknowledgment by the receiving operative in charge of the work
- signing off completed work
- cancellation of the permit by a senior manager (preferably the issuing manager).

Safety induction and training

The principal contractor should establish management procedures for safety awareness induction for all personnel who are appointed to the project site. Although the mode of delivery will differ between organisations, common elements of safety induction would include:

- specification of the organisation's health and safety policy and procedures
- communication of the construction-phase health and safety plan and safety method statements relating to particular aspects and characteristics of the works
- specification of expected safety performance standards
- detailing the penalties and disciplinary issues of breaching the procedures
- clarification of any project aspect and encouraging support and co-operation between site personnel.

Induction meetings will integrate the technical aspect of the work with the specified health and safety procedures. They will therefore:

- familiarise personnel with the construction requirements of the works;
- inform personnel about the safety aspects of undertaking those works; and
- determine supervisory management of both the technical and safety aspects.

The MHSWR 1992 require a designated company safety adviser/officer to be present at induction events where the works are complicated and assume high-risk characteristics. At such induction events, a typical agenda will require the safety adviser to:

- remind personnel of the core safety procedures and standards;
- identify task/operation-specific safety aspects;
- advise on designated access and egress points, traffic management, public welfare and safety, and site boundaries and restricted access;

- specify permit-to-work procedures;
- highlight first aid and emergency procedures;
- specify safety inspection arrangements and identify safety supervisors present on site;
- advise on disciplinary measures for non-compliance with procedures;
- specify modes of reporting safety incidents and near incidents; and
- provide personnel with a safety management manual.

Induction events shall be documented and attendance lists compiled. This will reinforce the basic message that safety issues are a serious matter and that the organisation is committed to good safety management. Further, it provides a record to follow up safety training events.

Safety training should be seen as an on-going management procedure with follow-up events being held where, for example, there is a major change to the construction programme, a variation to operational procedures, and new personnel taking-up post on site.

Project administration: site rules and management procedures

Management procedures set out the methods to be used and the minimal standard requirements for all general activities on the project site. While many aspects of general project administration are common across projects, the CDM Regulations require that site-specific safety rules be established which accommodate the unique requirements of the project. These are termed 'site rules'.

Management procedures may cover almost any aspect of project administration, but in general, procedures are set in place to cover those aspects of health and safety policy established in the construction-phase health and safety plan. These were detailed in the earlier section on health and safety management policies.

Site rules are project-specific and exist to provide safe systems of work where works are to take place in a high-risk environment or where the works themselves are hazardous to those carrying them out. Site rules will be issued in the site health and safety manual applicable to the project to ensure that all personnel are aware of and follow safe working procedures.

An example of notified site rules for working in a high-risk restricted area might be produced as follows:

Document Section G.1: Site rules for operating in Zone H (restricted area)
All personnel must:
1. register upon arrival and departure
2. wear safety helmets and non-spark footwear
3. follow safety procedures and signs at all times
4. be familiar with the use of fire handling equipment and use of breathing apparatus before entering area
5. work only within designated and cordoned areas
6. report any safety matter to Zone H plant safety supervisor.

Safety management audits

An important aspect of the principal contractor's health and safety management system is safety management auditing. An audit is

- a detailed review of how safety management is being applied;
- an analysis of the degree to which the system is being complied with;
- an appraisal to determine the benefits being gained from system implementation; and
- an identification of systems aspects where a change in procedure can make improvements to overall safety management.

A safety management audit may be conducted in-house in association with the organisation's corporate safety team or could be conducted by specialist consultants hired for the task. The approach of the audit should be to

- evaluate the current safety policy;
- approve the safe working procedures currently specified;
- appraise the construction-phase health and safety plan;
- identify the control mechanisms in place to ensure compliance with project requirements (identified above);
- review the current construction programme, site structure and organisation and management personnel responsible for safety aspects;
- appraise safety management reports of incidents and near incidents; and
- provide a detailed report on the findings of the audit.

Activities to be audited can cover almost any aspect of site procedures and activity. Common items that will be audited include standards of

- administration
- welfare
- accommodation
- access control
- waste management
- plant and materials management
- groundworks
- temporary work
- occupational health controls
- fire prevention
- hazardous substance protection
- public safety.

Further and detailed information on auditing for safety can be acquired from BS8750: 'Guide to Occupational Health and Safety Management Systems'.

Feedback of information to management and project file

The practicality and success of the health and safety management system established by the principal contractor is heavily dependent upon the efficacy of the information-gathering and recording mechanisms implemented. While different organisations will employ their own unique procedures for documenting activities for their health and safety management options, typical approaches use a series of pro formas to form a set of site safety management records.

The following list provides the basis of a set of site safety records, and each is illustrated by a systems pro forma which is self explanatory:

- level of management responsibilities (Figure 5.11)
- safety document register (Figure 5.12)
- principal contractor's risk assessment (Figure 5.13)
- safety method statement evaluation record (for appraising subcontractor's safety method statements) (Figure 5.14)
- safety construction record (integrating activities of subcontractors) (Figure 5.15)
- permit-to-work record (Figure 5.16)
- safe working procedures record (Figure 5.17)
- hazard/safety incident investigation record (Figure 5.18)
- accident report (Figure 5.19)
- site safety inspection report (Figure 5.20)
- safety induction/training record (Figure 5.21).

Referring back to Figure 5.10, it can be seen that the principal contractor's health and safety management system has two feedback loops. As with all management systems, information is fed back into safety policy making to monitor the proactive and synergistic ethos of the system, In addition, information is channelled to the project health and safety file, a requirement of the CDM Regulations.

Management Responsibilities		
Project:		
Document Ref. No.:		
Date issued:		Page of
Compiled by:	Signed:	Date:
Authorised by:	Signed:	Date:
Name	Function	Responsibilities

Fig. 5.11 Principal contractor's management responsibilities: suggested pro forma

Document Register					
Project:					
Document Reference No.	Title	Issue date	Amendment date	Amended reference	Authorised by

Fig. 5.12 Principal contractor's document register: suggested pro forma

Principal Contractor's Risk Assessment

Project:				Document Ref. No.:	
Contractor:				Specialist discipline:	
Assessor:			Signed:	Date:	

Activity/element	Potential hazards	Population at risk	Risk rating			Priority	Control measures specified
			L	S	R		

Sources of information:

Key:
L = Likelihood
S = Severity
R = Risk (Severity × Likelihood)

Fig. 5.13 Principal contractor's risk assessment: suggested pro forma

Principal Contractor's Safety Method Statement Review				
Project:				
Document Ref. No.:				
Document submitted by:				
Contractor:				
Specialist discipline:				
Evaluation of:				
The method statement is accepted*/returned for reconsideration* (*delete as applicable)				
Next action:				
Assessed by:				
Date:				

	TEST	YES	NO	IN PART	N/A
1	Task/process and area of specialisation				
2	Sequence of work				
3	Supervisory arrangements				
4	Monitoring arrangements				
5	Schedule of plant				
6	Reference to occupational health standards				
7	First aid				
8	Schedule for personal protective equipment				
9	Schedule of arrangements for demarcation				
10	Controls for the safety of third parties				
11	Are the assessed high-risk or safety-critical phases identified with controls specified?				
12	Emergency procedures				

Fig. 5.14 Principal contractor's safety method statement review: suggested pro forma

Principal Contractor's Site Safety Co-ordination Record											
Project:						Document Ref. No.:					
Activity	Contractor	Start date	Pre-qualification questionnaire		Initial safety meeting	Safety method statement evaluation	Risk assessment				Safety information given	Authorised to start
			Sent	Received and accepted			General	COSHH	Noise	Manual handling		

Fig. 5.15 Principal contractor's site safety co-ordination record: suggested pro forma

Principal Contractor's Permit to Work	
Project:	
Document Ref. No.:	
Task or Work Operation:	Duration of permit:
This permit to work is issued for the following:	
Is work to be carried out when plant, equipment or systems are in operation?	yes / no
Location of work:	
Description of work (specific hazards):	
Precautions to be taken:	
Extra precautions to be taken if plant and equipment are being used:	
Additional permits: ● Hot work ● Electrical ● Confined space ● Other	
Authorisation	
Name of person issuing permit	
Designation:	
Signature:	
Time:	Date:

Fig. 5.16 Principal contractor's permit to work: suggested pro forma (continued in next page)

Receipt	
Name:	
Designation:	
Signature:	
Company:	
Clearance	
The work stated above has/has not been completed. Details if not completed:	
Name:	
Designation:	
Signature:	
Company:	
Cancellation	
Permit to work is cancelled.	
Name:	
Designation:	
Signature:	
Date:	Time:

Fig. 5.16 Principal contractor's permit to work: suggested pro forma (cont'd)

Principal Contractor's Safe Working Procedures
Project:
Document Ref. No.:
Task or work operation:
This safe working procedure has been prepared for the following work.
Location of work:
Description of work:
Safe methods to be adopted:
Prepared by
Name:
Designation:
Signature:

Fig. 5.17 Principal contractor's safe working procedures: suggested pro forma

Principal Contractor's Incident Investigation Report			
Project:			
Document Ref. No.:			
Parties involved:			
Location of incident:			
Date of incident:		Time of incident: am/pm	
Type of incident			
Potential severity:	☐ Major	☐ Serious	☐ Minor
Probability of recurrence:	☐ High	☐ Medium	☐ Low
Description of how incident occurred:			
Immediate causes: what unsafe acts or conditions caused the event?			
Secondary causes: what human, organisational or job factors caused the event?			
Remedial actions: recommendations to prevent recurrence:			
Signature of investigator:		Date:	
Follow-up action/review of recommendations and progress:			
Name of reviewer:			
Position/title of reviewer:			
Signature of reviewer:		Date:	

Fig. 5.18 Principal contractor's safety incident investigation: suggested pro forma

Principal Contractor's Accident Report

Project:

Document Ref. No.:

Page of

Injured person:

Accident:

Person reporting accident:

Name:

Date: Time:

Name:

Home address:

Location:

Home address:

Work process involved:

Cause (if known):

Occupation:

Occupation:

Details of injury:

Signature:

Date of report:

Name:

Date: Time:

Name:

Home address:

Location:

Home address:

Work process involved:

Cause (if known):

Occupation:

Occupation:

Details of injury:

Signature:

Date of report:

Fig. 5.19 Principal contractor's accident report: suggested pro forma

Principal Contractor's Site Safety Inspection				
Project:				
Document Ref. No.:			Date:	Time:
Location:				
Any unsafe conditions or work:				
Remedial action:				
Further action to be taken:	By (named person)	Date	Complete	
Inspected by (safety inspector):				
Action authorised by:			Date:	

Fig. 5.20 Principal contractor's site safety inspection: suggested pro forma

Principal Contractor's Induction and Further Training		
Project:		
Document Ref. No.:		
Contractor:		
Type of training:		
Name of trainee	Trainer	Date of training

Fig. 5.21 Principal contractor's induction and further training: suggested pro forma

Health and safety management: example

This section presents examples of a pre-tender health and safety plan and a construction-phase health and safety plan.

Pre-tender health and safety plan

The project on which this example is based comprises minor civil and water engineering and major landscape works. The client for whom the project was carried out was a city council. The planning supervisor appointed was from a consulting services department. The principal designer incorporated landscape architects and urban designers, with civil engineering inputs provided by an engineering design sub-consulting practice.

The documentation shown is presented in four parts, as follows:

Part A: Designer appraisal forms.
Part B: Scheme design hazard assessment.
Part C: Designer's risk assessment.
Part D: Planning supervisor's pre-tender health and safety plan.

Construction-phase health and safety plan

The project on which this example is based was similar to but the not the same project as that used to illustrate the pre-tender health and safety plan.

The documentation shown is as follows:

Part E: Contractor's construction-phase health and safety plan.

In all parts, the names and addresses of principal participants have been deleted in the interests of confidentiality. The documents presented should be appreciated as an indication of approach. All health and safety documents and all associated documents will vary according to the nature of the individual project and the construction professionals involved.

Part A: Designer Appraisal

Designer Appraisal Form

SITE LOCATION

DESIGNER: Name and address of the organisation

	Designer Appraisal Form
A.	**Knowledge of construction practices and familiarity and knowledge of the design process:**
(1)	Provide details of your experience of acting as designer on projects of this nature:
(2)	Provide details of your qualifications and resume of your experience generally:
(3)	Provide details of any formal notices from the Health & Safety Executive or litigation with the Health & Safety Executive

	Designer Appraisal Form	
B.	**Knowledge of health & safety issues, particularly in preparing the Health & Safety Plan:**	
(1)	Provide details of your previous experience of projects incorporating health and safety issues:	
(2)	Provide details of your experience of compiling risk assessments:	
(3)	Provide details of your knowledge of the responsibilities of the client, the planning supervisor, principal contractor and designer in respect of the CDM Regulations	

	Designer Appraisal Form
C.	**Ability to work with and co-ordinate the activities of different designers**
(1)	Provide details of your experience of working with other consultants as part of the design team:
(2)	Provide details of your experience of working with building contractors on a variety of construction schemes:

Designers Appraisal Form

D.	**Number, experience and qualifications of the people to be employed (internal and external) to carry out the functions:**
(1)	Provide details of staff who are members of recognised construction bodies, related professional bodies, or hold a recognised health and safety qualification
(2)	Provide details of staff with experience as set out in section 1 above:
(3)	Provide details of staff who have received training in health and safety and who hold the relevant knowledge and experience

Designer Appraisal Form

E.	Management systems to carry out the role:
(1)	Provide details of a quality management system in force with procedure notes on each function prior to commencement of the works on site, for design, project management and health and safety services:

F.	Time allocation to carry out the different duties:
(1)	Provide details of time allowed to carry out the different duties:

G.	Technical facilities available:
(1)	Provide details of your health and safety library, if any:
(2)	Provide details of your commitment to and subscription to construction safety publications:

Designers Appraisal Form

H.	Communication systems:
(1)	Provide details of methods of communicating design decisions:
(2)	Provide details of methods of communicating other risks associated with the project:

SIGNATURE.. DATE......................

POSITION...

Part B: Scheme Design Hazard Assessment

SCHEME STAGE: SCHEME DESIGN

JOB NO:
COMPLETED BY:
CHECKED BY:

SHEET NO: 1
DATE:
DATE:

ASPECT OF WORK, CONSTRUCTION OPERATION OR COMPONENT	HAZARD IDENTIFIED	ACTION	Initial Ref. Referral	Action Reqd.	RISK IDENTIFIED	LEVEL	ACTION IDENTIFIED	Final Ref. Referral
EXISTING ENVIRONMENT								
Land uses and related restrictions								
Old river	Work/use of machinery on steep slopes	Reduce	H&S Plan	✓	Falling in	High	Consider requirement during detailed design	
	Use of cement, paint, etc.		H&S File		Pollution of watercourse	Med–low		
Adjacent development sites	Unsecured boundaries may contribute to vehicular/foot conflicts. Also implications for site security/trespass	Avoid	H&S Plan	✓	Personal injury	High	Provide hoardings to control access	
Services								
Underground street lighting cables	Electric shock	Reduce	H&S Plan	✓	Personal injury	Depends on voltage	Obtain service records. Contractor to verify/hand excavation in vicinity of cables	
Street lighting columns	Restrictions on site working and trafficking	Reduce	H&S Plan	✓	Personal injury	Medium		
Proposed sewer outfall from housing	Potential conflict with site works/constraint on site working	Reduce	H&S Plan	✓	Personal injury	Med–high	Consider phasing of work	
Traffic systems and restrictions								
(Public right of way)	Conflicts between site traffic and pedestrians/cyclists	Reduce	H&S Plan	✓	Personal injury	High	Erect hoardings; provide clear signage to control access	
	Conflicts between site traffic and pedestrians/cyclists	Reduce	H&S Plan	✓	Personal injury	High	Consider temporary closure. Erect hoardings; provide clear signage to control access	

SCHEME STAGE: SCHEME DESIGN

JOB NO:
COMPLETED BY:
CHECKED BY:

SHEET NO: 2
DATE:
DATE:

ASPECT OF WORK, CONSTRUCTION OPERATION OR COMPONENT	HAZARD IDENTIFIED	ACTION	Initial Ref. Referral	Action Reqd.	RISK IDENTIFIED	LEVEL	ACTION IDENTIFIED	Final Ref. Referral
Traffic systems and restrictions (cont.)								
	Conflict between traffic and pedestrians/cyclists at road junction/entrance to right of way and potential site access point	Reduce/ mitigate	H&S Plan	✓	Personal injury	High		
	Works by city council in vicinity of secondary site access	Reduce/ mitigate	H&S Plan	✓	Personal injury	High		
Adjacent developments	Potential stockpiling of fill materials on park area leading to site traffic conflicts	Remove	H&S Plan	✓	Personal injury	High		
Existing structures								
Former pump house	Falling objects/restriction to workspace	Mitigate	H&S Plan	✓	Personal injury	High		
Retaining wall	Collapse	Avoid	H&S Plan	✓	Personal injury	High	Design to incorporate any requirement for temporary support	
	Falling into river	Reduce	H&S Plan	✓	Personal injury	High		
	Escape of materials	Avoid	H&S Plan	✓	Pollution of watercourse	Med–low		
Former railway bridges	Falling objects	Mitigate	H&S Plan	✓	Personal injury	High		
	Fall from height							
	Falling objects	Mitigate	H&S Plan	✓	Personal injury	High		

215

SCHEME STAGE: SCHEME DESIGN

JOB NO:
COMPLETED BY:
CHECKED BY:

SHEET NO: 3
DATE:
DATE:

ASPECT OF WORK, CONSTRUCTION OPERATION OR COMPONENT	HAZARD IDENTIFIED	ACTION	Initial Ref. Referral	Action Reqd.	RISK IDENTIFIED	LEVEL	ACTION IDENTIFIED	Final Ref. Referral
Ground conditions								
Contaminated ground	Deep excavation for drainage or structural works below level of remediation	Mitigate	H&S Plan Specificn.	✓	Contact with substances hazardous to health	High		
Steep slopes generally and in proximity to water	Collapse or falling: falling into river	Mitigate	H&S Plan H&S File	✓	Personal injury	High		
Non-reinstatement of pre-remediation site levels	Falling or tripping	Mitigate	H&S Plan	✓	Personal injury	Medium		
	Exposure of underground supply cables (street lighting)				Electric shock	High		
SITE-WIDE ELEMENTS								
Positioning of site access/egress point for deliveries and emergencies	Potential conflict with traffic on public highway pedestrians/cyclists on designated cycleway	Avoid/mitigate	H&S Plan	✓	Personal injury	High		
Traffic and/or pedestrian routes	Conflict with site traffic	Avoid/mitigate	H&S Plan	✓	Personal injury	High		

SCHEME STAGE: SCHEME DESIGN

JOB NO:
COMPLETED BY:
CHECKED BY:

SHEET NO: 4
DATE:
DATE:

ASPECT OF WORK, CONSTRUCTION OPERATION OR COMPONENT	HAZARD IDENTIFIED	ACTION	Initial Ref. Referral	Action Reqd.	RISK IDENTIFIED	LEVEL	ACTION IDENTIFIED	Final Ref. Referral
HAZARDS OR WORK SEQUENCES WHICH MAY ARISE AS A RESULT OF DESIGN								
Earthmoving/mounding	Use of heavy machinery	Mitigate	H&S Plan	✓	Personal injury	High		
Drainage works	Unprotected deep excavations	Mitigate	H&S Plan	✓	Falling/personal injury Collapse	High		
	Work in confined spaces	Mitigate	H&S Plan	✓	Collapse; inadequate ventilation	High		
	Work below remediated layer	Avoid/ mitigate	H&S Plan	✓	Contact with substances harardous to health	Depends on substance		
Built features including walls, piers, focal features	Lifting of heavy units	Reduce	Specificn.	✓	Personal injury	Depends on object	Reduce size or incorporate lifting sockets	
Construction of footpath and associated works to river bank	Use of machinery on steep slope above river	Reduce	H&S Plan	✓	Falling/personal injury	High		
	Work within 8m of river bank	Reduce	H&S Plan	✓	Falling/personal injury	High		
	Spillage of materials	Avoid	H&S Plan	✓	Pollution of watercourse	Med–low		
Works to embankment to disused railway bridge	Use of machinery on steep slope	Reduce	H&S Plan	✓	Falling/personal injury	High		
Construction of new retaining wall to river	Work in proximity to watercourse	Mitigate	H&S Plan	✓	Falling/personal injury Collapse Pollution of watercourse	High Med–low	Design to take account of any temporary supports required	

SCHEME STAGE: SCHEME DESIGN

JOB NO:
COMPLETED BY:
CHECKED BY:

SHEET NO: 5
DATE:
DATE:

ASPECT OF WORK, CONSTRUCTION OPERATION OR COMPONENT	HAZARD IDENTIFIED	ACTION	Initial Ref. Referral	Action Reqd.	RISK IDENTIFIED	LEVEL	ACTION IDENTIFIED	Final Ref. Referral
HAZARDS OR WORK SEQUENCES WHICH MAY ARISE AS A RESULT OF DESIGN (cont.)								
Maintenance of grassed and planted areas to river bank	Use of machinery on steep slopes	Reduce			Falling/personal injury	High	Reduced mowing frequency by use of wild flora/grass seed mix	
	Spillage of contaminants e.g. diesel	Avoid	H&S Plan H&S File	✓	Pollution of watercourse	Med–low	Restriction of mowing to visible areas accessible from paths. Maintenance by strimming in place of herbicides or use of lower-risk herbicides	
Maintenance of softworks to embankment to railway bridge	Use of machinery on steep slopes	Reduce	H&S Plan H&S File	✓	Falling/personal injury	High	Reduced mowing frequency by use of wild flora/grass seed mix. Restriction of mowing to visible areas accessible from paths	

SCHEME STAGE: SCHEME DESIGN

JOB NO:
COMPLETED BY:
CHECKED BY:

SHEET NO: 6
DATE:
DATE:

ASPECT OF WORK, CONSTRUCTION OPERATION OR COMPONENT	HAZARD IDENTIFIED	ACTION	Initial Ref. Referral	Action Reqd.	RISK IDENTIFIED	LEVEL	ACTION IDENTIFIED	Final Ref. Referral
CERTAIN ASPECTS OF DESIGN HAVING ON-GOING IMPLICATIONS FOR PUBLIC SAFETY, for example								
Shared footway/cycleways	Conflict between pedestrians and cyclists	Reduce	H&S Plan	✓	Personal injury	Med–high	Provide appropriate signage	
Requirement for EVA/maintenance access/service access to green	Potential (occasional) conflict between pedestrians/cyclists and vehicles		H&S File	✓	Risk of falling in/ drowning	Med–high		
Waterside access	Open water	Reduce	H&S File	✓	Risk of falling in/ drowning	Med–high	Treatment in accordance with R.O.S.P.A. banding e.g. Encourage natural growth separation margins; access route signage of hazard	
Waterside access: elevated	Open water	Reduce	H&S File	✓	Risk of falling in/ drowning	Med–high	Treatment in accordance with R.O.S.P.A. banding e.g. 'denial of access' balustrade; national signage display of hazard	
Maintenance of planted areas	Use of herbicides	Mitigate	H&S File	✓	Spray drift/contact	Low–med	Ensure application by suitably qualified personnel in accordance with regulations	

SCHEME STAGE: SCHEME DESIGN

JOB NO:
COMPLETED BY:
CHECKED BY:

SHEET NO: 7
DATE:
DATE:

CONSTRUCTION MATERIALS – SPECIFICATION OR DESIGN ASPECTS

ASPECT OF WORK, CONSTRUCTION OPERATION OR COMPONENT	HAZARD IDENTIFIED	ACTION	Initial Ref. Referral	Action Reqd.	RISK IDENTIFIED	LEVEL	ACTION IDENTIFIED	Final Ref. Referral
Paving type and pattern	Requirement for cutting Requirement for lifting	Reduce/ mitigate	Specificn.	✓	Use of saw or other cutting equipment Generation of dust or flying debris Lifting of heavy objects	Med–high	Co-ordination of setting out dimensions with unit size to minimise need for cutting Consider selection of paving material (size/type) and bond/bond angle	
Heavy elements e.g. copings to walls and feature piers; precast concrete units; sculptural features; railway sleeper steps	Requirement for lifting	Reduce/ mitigate	H&S File Specificn.	✓	Lifting of heavy objects	High	Selection of alternative, lighter materials; or reduced unit size Incorporation of sockets for mechanical lifting	
Metalwork to fencing, furniture, artwork, etc.	Cutting and welding of components Applied finishes Requirement for lifting	Reduce/ mitigate	H&S Plan Specificn.	✓	Contact with machinery or substances hazardous to health by skin contact or inhalation	Depends on substance	Site fabrication v. pre-made Minimisation of site cutting/welding Off-site treatments under controlled environment e.g. galvanising to minimise preparation; pre-priming Number and weight of sections	

SCHEME STAGE: SCHEME DESIGN

JOB NO:
COMPLETED BY:
CHECKED BY:

SHEET NO: 8
DATE:
DATE:

ASPECT OF WORK, CONSTRUCTION OPERATION OR COMPONENT	HAZARD IDENTIFIED	ACTION	Initial Ref. Referral	Action Reqd.	RISK IDENTIFIED	LEVEL	ACTION IDENTIFIED	Final Ref. Referral
CONSTRUCTION MATERIALS – SPECIFICATION OR DESIGN ASPECTS								
Applied coatings and finishes to timber and metal components	Use of primers, paints, thinners, preservative stains, etc.	Reduce/ mitigate	H&S Plan Specificn.	✓	Contact with substances hazardous to health by skin contact or inhalation	Depends on substance	Off-site galvanising of metalwork to minimise preparation. Off-site priming to minimise site painting. Selection of finishes or paints less toxic in nature. Untreated finishes e.g. avoidance of 'cosmetic' stains to preservative-treated timber?	

SCHEME STAGE: SCHEME DESIGN

JOB NO:
COMPLETED BY:
CHECKED BY:

SHEET NO: 9
DATE:
DATE:

ASPECT OF WORK, CONSTRUCTION OPERATION OR COMPONENT	HAZARD IDENTIFIED	ACTION	Initial Ref. Referral	Action Reqd.	RISK IDENTIFIED	LEVEL	ACTION IDENTIFIED	Final Ref. Referral
ASPECTS OF CONSTRUCTION WORKS WITH INHERENT HAZARDS WHICH ARE UNLIKELY TO BE AVOIDED AND NEED TO BE COVERED IN DETAIL BY PRINCIPAL CONTRACTOR'S SAFETY PLAN								
• Required continuance of public access along right of way • Work in the public highway and earthmoving operations • Possibility of contaminants below level of remediation • Excavations greater than 1.2m deep • Buried services, live and abandoned • Confined spaces in manholes • Overhead working (on adjacent sites) • Erection of building components using cranes (on adjacent sites) • Stability of partially complete structures • Unprotected edges/openings • Stability of existing structures • Connections to live services (electricity, water, sewerage) • Proximity of existing buildings/structures or adjacent development sites • Works to existing trees • Temporary works to excavations and concrete formwork • Carrying, lifting and placing of structural elements • Any hotwork/cutting/grinding/welding associated with the work • Work to abutments, retaining walls and river banks • Heavy plant working on, turning or reversing near sloping river bank • Use of herbicides • Noise and vibration • Fire **MAINTENANCE HAZARDS** • Manholes and inspection chambers • Work near or adjacent to river's edge • Work on steep slopes, including near to river's edge • Proximity of water to the general public • Shared use of footway/cycleway • Use and disposal of chemicals, e.g. herbicides • Servicing of events on green/EVA/maintenance access								

SCHEME STAGE: SCHEME DESIGN

JOB NO:
COMPLETED BY:
CHECKED BY:

SHEET NO: 10
DATE:
DATE:

ASPECT OF WORK, CONSTRUCTION OPERATION OR COMPONENT	HAZARD IDENTIFIED	ACTION	Initial Ref. Referral	Action Reqd.	RISK IDENTIFIED	LEVEL	ACTION IDENTIFIED	Final Ref. Referral

ASPECTS OF CONSTRUCTION WORKS WITH INHERENT HAZARDS WHICH ARE UNLIKELY TO BE AVOIDED AND NEED TO BE COVERED IN DETAIL BY PRINCIPAL CONTRACTOR'S SAFETY PLAN (cont.)

HEALTH HAZARDS

Examples of substances harmful by inhalation:

- Welding fumes
- Hardwood dust
- Cement dust
- Isocyanates (paints, varnishes, adhesives)
- Solvents (paints, paint strippers, mastics, glues, surface coatings)

Examples of substances hazardous in contract with skin and mucous membranes:

- Bitumen
- Brick, concrete, stone dust
- Cement
- Paints, varnishes, stains
- Certain epoxy resins
- Chromates (in primer paints, cement)
- Petrol, white spirit, thinners
- Acids
- Alkalis

Part C: Designer's Risk Assessment

DESIGNER'S RISK ASSESSMENT				JOB NO: COMPLETED BY: CHECKED BY:	SHEET NO: 1 DATE: DATE:
ASPECT OF WORK	**POTENTIAL HAZARD**	**AT RISK**	**LEVEL**	**POSSIBLE CONTROL MEASURES**	**ACTION BY**
Excavations below level of remediation	Exposure to contaminants	C	Low	Reference to be made to completion report for remediation contract. Contractor to prepare method statement for safe working	Main contractor
Works in proximity to watercourse	Personal injury Release of materials Restriction on site working	C	Medium	Care to be taken while working in vicinity. Work to be undertaken by suitably trained and qualified operatives. Contractor to prepare method statement for safe working	Main contractor
Work on steep slopes	Personal injury Restriction on site working	C	Low	Care to be taken while carrying out works. Work to be undertaken by suitably trained and qualified operatives. Appropriate plant to be used when working on slopes	Main contractor/ landscape subcontractor
Existing services	Personal injury Restriction on site working	C	Low	Care to be taken when working in vicinity. Position of services to be verified on site with relevant utilities	Main contractor/ landscape subcontractor
Public right of way Public access retained until opening of final route	Conflict between public and site traffic Restriction on site working	C.V.P.	Medium	Particular care to be taken when working adjacent to pedestrian/cycle traffic. All routing of site traffic, signing and barriers to be in accordance with safe practice and local authority requirement. Contractor to prepare method statement for phasing and safe working	Main contractor
Public access retained until sectional completion	Conflict between public and site traffic Restriction on site working	C.V.P.	Medium	Stopping up order required for any essential temporary closure. Particular care to be taken when working adjacent to pedestrian/cycle traffic. All routing of site traffic, signing and barriers to be in accordance with safe practice and local authority requirement. Contractor to prepare method statement for phasing and safe working	Main contractor

NOTES:

1. At risk. C = contractor, V = visitors, P = public.
2. Level. Low = normal safety precautions required; Medium = special safety precautions to be considered; High = special safety precautions essential.
3. Hazards to be considered: slips, trips and falls, entrapment, drowning, flying/falling materials, collapse, handling, lifting, overturning, overloading, fire & explosion, noise, dust, vibration, fumes, harmful substances, contamination, radiation, disease, cuts, injury, traffic, access, temporary works, confined spaces, deep excavations, overhead & underground working, live services, machinery, etc.

DESIGNER'S RISK ASSESSMENT

JOB NO:
COMPLETED BY:
CHECKED BY:

SHEET NO: 2
DATE:
DATE:

ASPECT OF WORK	POTENTIAL HAZARD	AT RISK	LEVEL	POSSIBLE CONTROL MEASURES	ACTION BY
Public access retained for duration of works	Conflict between public and site traffic Restriction on site working	C.V.P.	Medium	Stopping up order required for any essential temporary closure. Particular care to be taken when working adjacent to pedestrian/cycle traffic. All routing of site traffic, signing and barriers to be in accordance with safe practice and local authority requirement. Contractor to prepare method statement for phasing and safe working	Main contractor
Concurrent works on adjacent sites	Conflict with other site traffic/personnel Restrictions on site working	C	Low	Care to be taken when working in vicinity. Contractor to prepare method statement for site working and allow for all effects of any co-ordination required with adjacent site works	Main contractor/specialist subcontractor
Works by others within site area – outfall to housing development	Conflict between 2 sets site traffic Restriction on working/programming	C	Low	Contractor to allow for access to site and all effects/co-ordination required in programming works. Contractor to prepare method statement for safe working	Main contractor
'Artists and Tradesman' works within site area – installation of play equipment	Conflict between 2 sets site traffic Restriction on working/programming	C	Low	Contractor to allow for all effects and co-ordination required in programming works. Contractor to prepare method statement for safe working	Main contractor/specialist subcontractors
Site access off public highway	Conflict with traffic/pedestrians	C.V.P.	Low	Particular care to be taken when working adjacent to public highways. Traffic management in accordance with highways requirements by suitably trained and qualified operatives	Main contractor
	Conflict with traffic/restriction on site working area	C	Low	Care to be taken when working in area. Work to be undertaken by suitably trained and qualified operatives	Main contractor

NOTES:

1. At risk. C = contractor, V = visitors, P = public.
2. Level. Low = normal safety precautions required; Medium = special safety precautions to be considered; High = special safety precautions essential.
3. Hazards to be considered: slips, trips and falls, entrapment, drowning, flying/falling materials, collapse, handling, lifting, overturning, overloading, fire & explosion, noise, dust, vibration, fumes, harmful substances, contamination, radiation, disease, cuts, injury, traffic, access, temporary works, confined spaces, deep excavations, overhead & underground working, live services, machinery, etc.

DESIGNER'S RISK ASSESSMENT

JOB NO:
COMPLETED BY:
CHECKED BY:

SHEET NO: 3
DATE:
DATE:

ASPECT OF WORK	POTENTIAL HAZARD	AT RISK	LEVEL	POSSIBLE CONTROL MEASURES	ACTION BY
Subsoil stockpile	Possible double handling leading to increased risk of site traffic conflicts; restrictions on site working	C	Low	Work by suitably trained and qualified operatives. Contractor to allow for all effects in programming works and prepare method statement for safe working, identifying any temporary stockpile areas required	Main contractor
Topsoil storage area	Site traffic conflicts/conflicts with public	C.V.P.	Medium	Work by suitably trained and qualified operatives. Direct access to site only to be allowed in advance of sectional completion and opening to public. Contractor to prepare method statement for safe working, identifying proposed phasing of works and haulage routes.	Main contractor
Site boundaries with adjacent developments	Vehicular/foot conflicts to unsecured boundaries. Security/trespass	C	Medium	Contractor to prepare method statement for safe working, identifying temporary fencing or signage required in advance of construction of permanent boundaries	Main contractor
Disused building	Falling objects / Restricted work space	C.V.P.	Low	Contractor to prepare method stagement and give particular care to safe woking in vicinity. Work by suitably trained and qualified operatives. Erection of hoarding to preclude public access prior to opening of park	Main contractor
Existing bridges over river adjacent to site boundary	Falling objects / Fall from height	C.V.P.	Low	Outwith site works area. Care to be taken when working adjacent to bridges work by suitability trained and qualified operatives	Main contractor
Maintenance of soft landscape works	Conflict with public (vehicles, chemicals, etc.)	C.P.	Low	Particular care to be taken when working in areas open to public. Works to be carried out by suitably trained and qualified operatives in accordance with all current regulations enactments	Main contractor/ landscape subcontractor

NOTES:

1. At risk. C = contractor, V = visitors, P = public.
2. Level. Low = normal safety precautions required; Medium = special safety precautions to be considered; High = special safety precautions essential.
3. Hazards to be considered: slips, trips and falls, entrapment, drowning, flying/falling materials, collapse, handling, lifting, overturning, overloading, fire & explosion, noise, dust, vibration, fumes, harmful substances, contamination, radiation, disease, cuts, injury, traffic, access, temporary works, confined spaces, deep excavations, overhead & underground working, live services, machinery, etc.

DESIGNER'S RISK ASSESSMENT

JOB NO:
COMPLETED BY:
CHECKED BY:

SHEET NO: 4
DATE:
DATE:

ASPECTS OF CONSTRUCTION WORKS WITH INHERENT HAZARDS WHICH ARE UNLIKELY TO BE AVOIDED AND NEED TO BE COVERED IN DETAIL BY PRINCIPAL CONTRACTOR'S SAFETY PLAN

- Required continuance of public access along right of way
- Work in the public highway and earthmoving operations
- Possibility of contaminants below level of remediation
- Excavations greater than 1.2m deep
- Buried services, live and abandoned
- Confined spaces in manholes
- Overhead working (on adjacent sites)
- Erection of building components using cranes (on adjacent sites)
- Stability of partially completed structures
- Unprotected edges/openings
- Stability of existing structures
- Connections to live services (electricity, water, sewerage)
- Proximity of existing buildings/structures or adjacent development sites
- Works to existing trees
- Temporary works to excavations and concrete formwork
- Carrying, lifting and placing of structural elements
- Any hotwork/cutting/grinding/welding associated with the work
- Work to abutments, retaining walls and river banks
- Heavy plant working on, turning or reversing near sloping river bank
- Use of herbicides
- Noise and vibration
- Fire

MAINTENANCE HAZARDS

- Manholes and inspection chambers
- Work near or adjacent to river's edge
- Work on steep slopes, including near to river's edge
- Proximity of water to the general public
- Shared use of footway/cycleway
- Use and disposal of chemicals, e.g. herbicides
- Servicing of events on green/EVA/maintenance access

DESIGNER'S RISK ASSESSMENT

JOB NO:
COMPLETED BY:
CHECKED BY:

SHEET NO: 5
DATE:
DATE:

ASPECTS OF CONSTRUCTION WORKS WITH INHERENT HAZARDS WHICH ARE UNLIKELY TO BE AVOIDED AND NEED TO BE COVERED IN DETAIL BY PRINCIPAL CONTRACTOR'S SAFETY PLAN (cont.)

Examples of substances harmful by inhalation:

- Welding fumes
- Hardwood dust
- Cement dust
- Isocyanates (paints, varnishes, adhesives)
- Solvents (paint strippers, mastics, glues, surface coatings)

Examples of substances hazardous in contact with skin and mucous membranes:

- Bitumen
- Brick, concrete, stone dust
- Cement
- Paints, varnishes, stains
- Certain epoxy resins
- Chromates (in primer paints, cement)
- Petrol, white spirit, thinners
- Acids
- Alkalis

Part D: Planning Supervisor's Pre-Tender Health & Safety Plan

CONTENTS

1. INTRODUCTION 1
   1.1 Generally 1
   1.2 Method statements 1
   1.3 Health and Safety File 1
   1.4 Questionnaire 1

2. NATURE OF THE PROJECT 2
   2.1 The client 2
   2.2 Client's agent 2
   2.3 Location 2
   2.4 Nature of the works 2
   2.5 Time-scale 2

3. THE EXISTING ENVIRONMENT 3
   3.1 Site description 3
   3.2 Existing land, ground conditions and nature of soil 3
   3.3 Groundwater level 3
   3.4 Existing structures 3
   3.5 Surrounding land uses 4
   3.6 Proposed adjacent development 4
   3.7 Restricted access 4
   3.8 Existing services 4
   3.9 Traffic systems and restrictions 5
   3.10 Health hazards 5
        3.10.1 Asbestos 5
        3.10.2 Hazardous material generally 6

4. EXISTING DRAWINGS AND DOCUMENTS 7
   4.1 Health and Safety File 7
   4.2 Drawings included in the Health and Safety File 7
   4.3 Other documents included in the Health and Safety File 8
   4.4 Existing drawings included as part of the tender documents 8

5. THE DESIGN 9
   5.1 Working over water 9
        5.1.1 Falls from heights 9
        5.1.2 Drowning 9
        5.1.3 Disease 9
   5.2 Gabions 10
   5.3 Excavations below level of remediation 10
   5.4 General excavations greater than 1.2m deep 10
   5.5 Drainage work 10
   5.6 Connections to existing services 10
   5.7 Work on steep slopes 11
   5.8 Street lighting installations 11
        5.8.1 Electrocution and fire 11
        5.8.2 Risks to the public 11
        5.8.3 Falling/tripping 11
        5.8.4 General site activities/contractor's staff 11
        5.8.5 Manual handling of heavy equipment 11
   5.9 Feature screen to plaza housing access features 12
        5.9.1 Manual handling of heavy equipment 12
        5.9.2 On-site welding operations 12

5.10 General manual handling 12
5.11 Work to existing trees 12
5.12 Maintenance of soft landscaped areas 12

6. CONSTRUCTION MATERIALS **13**
6.1 Construction materials generally 13
    Examples of substances harmful by inhalation 13
    Examples of substances hazardous in contact with skin and mucous
    membranes 13
6.2 Hot bituminous material 13

7. SITE-WIDE ELEMENTS **14**
7.1 Site access and egress points 14
7.2 Movement of site traffic 14
7.3 Location of temporary site accommodation 14
7.4 Location of unloading, layout and storage areas 14
7.5 Traffic/pedestrian routes 14
7.6 Site security 15
7.7 Co-ordination with work to be executed by others 15
    7.7.1 Installation of play equipment 15
    7.7.2 Outfall to housing development 15

8. OVERLAP WITH CLIENT'S UNDERTAKING **16**
8.1 Buildings in occupation 16

9. SITE RULES **17**
9.1 Generally 17
9.2 Conformity with all statutory requirements 17
9.3 Booking in/out 17
9.4 Subsoil stockpile 17
9.5 Topsoil storage area 17
9.6 Control of movements 17
9.7 'Absence' procedure 18
9.8 Unforeseen eventualities 18
9.9 Fires on site 18
9.10 Site security 18
9.11 Access around the site 18
9.12 Wheel wash facility 18
9.13 Provision of welfare facilities 18
9.14 First aid 19
9.15 Personnel on site 19
9.16 Communication around the site 19
9.17 Control of noise 19
9.18 Control of pollution 19
9.19 Training 19
9.20 Competence of site personnel 20

10. CONTINUING LIAISON **21**
10.1 Generally 21
10.2 Recording health and safety matters 21

11. TENDER STAGE METHOD STATEMENTS **22**

1.   INTRODUCTION

1.1   Generally

All prospective principal contractors tendering for this contract will receive this health and safety plan. The purpose is to highlight the main health and safety issues in connection with the construction work on the project and to form a basis for tenderers to explain their proposals for managing the problems.

The principal contractor will develop this health and safety plan, in particular taking reasonable steps to ensure co-operation between all contractors to achieve compliance with the rules and recommendations of the plan.

1.2   Method statements

Detailed method statements, produced by the principal contractor before work starts, are essential for safe working and are a requirement of Section 2(2) of the Health and Safety at Work Act. They should identify problems and their solutions and form a reference for site supervision. The method statements should be easy to understand, should be agreed by and known to all levels of management and supervision and should include such matters as outlined in this document.

1.3   Health and Safety File

A HEALTH AND SAFETY FILE for the site is available for inspection by appointment during normal office hours.

A copy of the health and safety plan will be handed to the principal contractor for use during the works and for developing for final handover to the client.

1.4   Questionnaire

Each tenderer is required to complete a questionnaire outlining their experiences of health and safety on site. This questionnaire is provided as a separate document and a copy is attached to this document as Appendix C. The questionnaire is to be completed and returned to the address below no later than one week before the date for return of tenders.

2.   NATURE OF THE PROJECT

2.1   The client

2.2   Client's agent

2.3   Location

2.4   Nature of the works

    2.4.1   Generally

        Hard and soft landscape works on remediated land together with works to the riverside.

    2.4.2   General landscape works

        – Hard and soft landscaped areas
        – Footpaths
        – Planting trees and shrubs
        – Boundary railings

- Site furniture
- Creation of plaza including feature screen
- Creation of landscaped mounds

2.4.3 Civil engineering work

- Foundations and sub-bases
- Riverside walkways
- Walls and piers
- Work to retaining wall
- Land drainage and other drainage
- Head walls

2.4.4 Play area and equipment

- Construction of bark pits
- Construction of drainage sumps
- Landscaping
- Play equipment and site furniture
- Fencing

2.4.5 Street lighting

- Lighting installations to new and existing pedestrian routes

2.5 Time-scale

Contract period
The contractor is to refer to the preliminaries/general conditions for details of contract dates and for sectional completion

3. THE EXISTING ENVIRONMENT

3.1 Site description

The site occupies a flat, low-lying area situated between the river and the canal to the west. The river and canal are joined to the north and south of the site, so forming an island. The site occupies a total area of approximately 15 hectares.

A pedestrian route runs across the site from east to west. The contractor is to allow for maintaining this footpath and for ensuring the safety of all pedestrians using it.

The site is currently secured by a 2.4 metre high hoarding around the perimeter of the site and between the park and the proposed residential area. There is no hoarding between the park and the proposed business park.

3.2 Existing land, ground conditions and nature of soil

Former land uses included warehousing, metal recycling (or 'scrap yards'), railway goods and maintenance yard and a power station. Ground contaminations were identified arising out of these previous site uses. All such materials have since been removed by specialists as part of a separate reclamation contract.

The base of reclamation excavations for the proposed park area was generally 58.00 AOD. At the entrance the base of reclamation excavations was lowered to 56.00 AOD. The 'as constructed' reclamation finished levels provided by the reclamation contractor are presented on drawings in the tender documents.

Full details of the reclamation contract are included in the Health and Safety File.

3.3 **Groundwater level**

The groundwater level on the site is described in the trade preambles for excavations and hard landscape work.

3.4 **Existing structures**

Prior to reclamation, all buildings, except the 'machine house', were demolished. The 'machine house' comprises a brick-built structure and has two main chambers, accessed from the north and south side of the building, respectively. It also has a tower, which was believed to house a steam-driven hydraulic pump.

While work to this structure does not form part of this contract, work will be going on around and adjacent to it and the principal contractor should make themselves aware of its condition.

3.5 **Surrounding land uses**

There are buildings close by, therefore the principal contractor can expect busy pedestrian traffic.

This pedestrian route will remain open at all times. This is a temporary footpath which will be relocated under this contract.

The cycle path, running north to south along the western side of the site, is also to remain open at all times. Part of this path is to be resurfaced and part is to be re-routed under this contract.

The principal contractor shall adopt suitable working methods to ensure the safety of all members of the public on these pedestrian and cycle routes.

3.6 **Proposed adjacent development**

To the north and south of the park are areas of proposed development.

These proposed developments comprise a business park to the north and a residential area to the south. The developer of the residential area is currently on site and the contractor is to assume that the developers for both that area and the science park will be on site during the course of the park contract.

The principal contractor is to take into account the possibility of:

– overhead working on the adjacent site;
– the erection of building components using cranes on adjacent sites.

The principal contractor is to detail any temporary fencing and signage to be used in advance of the construction of permanent boundaries.

3.7 **Restricted access**

There is restricted access to parts of the site due to neighbouring properties. Care is to be taken while working in these areas, and the entrances to properties must remain open at all times.

3.8 **Existing services**

The extent of known site services information is indicated on drawings and also on various drawings in the Health and Safety File. Where positions of existing services are indicated on the drawings they are for guidance only; it is the responsibility of the contractor to obtain details and exact positions from relevant service authorities.

Existing services in the public footpaths along routes across the site will remain live. Alterations to the street lighting installations will form part of the works.

Considering the depth of remediations it is not anticipated that any other underground services will be encountered in normal excavations. However, the contractor is to proceed with caution during all excavations, particularly during excavations below the remediation level, and is to establish formal procedures to ensure that any unidentified services located during the works are carefully checked to determine if they are live or contain any hazardous material or substances. All details must be recorded. The services engineer and the planning supervisor are to be provided with copies of the records setting out the nature and location of all such services prior to the agreement of a course of action.

3.9 Traffic systems and restrictions

The principal contractor is to determine and apply all local authority and police regulations.

3.10 Health hazards

3.10.1 Asbestos

Asbestos may be present around the 'machine house'. The removal of any material containing asbestos must be carried out strictly in accordance with the following statutory measures:

(a) Health and Safety at Work etc. Act 1974
(b) Control of Pollution Act 1974
(c) Control of Pollution (Special Waste) Regulations 1980
(d) Asbestos (Licensing) Regulations 1983
(e) Control of Asbestos at Work Regulations 1987
(f) any other legislation that may be applicable

The principal contractors and its subcontractors are responsible for the safety, health and welfare of their 'own employees'.

Attention is also drawn in particular to the requirements of the 1983 Regulations:

(a) for all operations to be carried out by a licensed contractor;
(b) for 28 days notice, or such shorter period as may be agreed, to be given to the relevant enforcement authority, whatever type of asbestos is involved.

If in doubt about the identity of the relevant Enforcement Authority the Contractor should contact the HSE.

3.10.2 Hazardous material generally

Hazardous material present within the site area, particularly on the embankments of the river, may include broken glass and used syringes among other domestic debris. Suitable precautions should be taken to prevent injury to the workforce or members of the public and all broken glass and dangerous objects removed from site.

4.   **EXISTING DRAWINGS AND DOCUMENTS**

4.1   **Health and Safety File**

A HEALTH AND SAFETY FILE for the site is available for inspection by appointment during normal office hours.

A copy of the health and safety plan will be handed to the principal contractor for use during the works and for developing for final handover to the client.

4.2   **Drawings included in the Health and Safety File**

List as required

4.3   **Other documents included in the Health and Safety File**

List as required

4.4   **Existing drawings included as part of the tender documents**

List as required

5.   **THE DESIGN**

The following principal hazards or work sequences so far identified cannot be avoided and will be a risk to health and safety of construction workers and/or the occupants of, or visitors to, the site.

It will be the responsibility of the principal contractor to detail their proposals for managing these problems. These details/method statements will be incorporated into the health and safety plan prior to the work commencing on site.

5.1   **Working over water**

Care is to be taken while working adjacent to the watercourses. The contractor should carefully consider the method of working and material-handling procedures. Construction operatives must be suitably trained and qualified. Proper equipment such as scaffolds, platforms, safety nets, safety belts, harnesses and lanyards should be considered. Suitable emergency procedures must be used such as the use of lifejackets, buoyancy aids and rescue lines, together with the information and training necessary to use the equipment. The construction operatives must be capable of responding effectively to an emergency.

The main hazards associated with working over water are:

− falls from heights
− drowning
− disease

5.1.1   **Falls from heights**

The contractor is to submit details of how to prevent personnel and debris from falling into the watercourses.

5.1.2   **Drowning**

All personnel working in the vicinity of the watercourses must wear buoyancy aids.

In the event of a person falling into the water two things are of paramount importance:

– the person must be kept afloat;
– location and rescue must be achieved as quickly as possible.

The contractor is to submit details for achieving these aims, including details of buoyancy aids and rescue equipment, if any, to be employed and details of rescue procedures. In addition, details are required of training given to personnel in order to effect such rescues.

5.1.3 Disease

With work being carried out adjacent to the waterways, the contractor's attention is drawn to the possible presence of vermin and all the associated health hazards.

For further information, the contractor is recommended to read the CITB Construction Site Safety Note 30 'Working Over Water'.

A document titled 'Special Requirements in Relation to the National Rivers Authority' is included as Appendix B. The contractor is to comply with all such requirements.

The installation of gabions along the edge of a section of the river is required.

As well as all the risks associated with working over water as identified in item 5.1, the installation of gabions poses other problems, including:

– proximity to the old railway bridge;
– limited headroom.

The principal contractor is to submit a method statement demonstrating how the installation of the gabions will be achieved.

5.3 Excavations below level of remediation

Excavations below the base of the reclamation excavations will generate material not tested as part of the reclamation.

5.4 General excavations greater than 1.2m deep

Most excavations will be in made-up ground. The contractor's method statements must identify proposals for:

– the prevention of collapse;
– the protection of edges of excavations to prevent falls of personnel, plant or materials into the trench.

5.5 Drainage work

The main problems associated with the proposed drainage work will be:

– deep excavations (including work below remediation level);
– work in confined spaces.

5.6 Connections to existing services

The contractor's proposals are required for safe working methods for connections to the following live services:

- electricity
- water
- sewerage

5.7 Work on steep slopes

Work on steep slopes includes landscape works to mounds and the clearing of rubbish and debris to the banks of the river. The main hazards associated with working on steep slopes include:

- angle of slope
- slippery surfaces
- hidden holes or obstacles
- visibility
- stability of machinery on bank
- position of body while carrying out the work

Care is to be taken when carrying out these works and the work is to be undertaken by suitably trained and qualified operatives wearing suitable protective clothing. Appropriate plant is to be used.

5.8 Street lighting installations

The work includes the disconnection and removal of existing street lighting columns along the pedestrian routes and the installation of new cabling and columns. The existing cabling is to remain in the ground; the contractor is to ensure that it is disconnected and made safe.

The risks associated with the street lighting installations include:

5.8.1 Electrocution and fire

These risks may be avoided by compliance with:

1. Electricity at Work Regulations
2. IEE Wiring Regulations
3. Health and Safety at Work Act
4. Managing Health and Safety on Construction Sites
5. Electrical Specification

No installation is to be energised until installation is complete, inspected and tested and a compliant 'Electrical Installation Certificate' is received.

All trades are to be made aware of the hazards and adequate protection is to be provided to safeguard staff and to prevent unauthorised entry.

5.8.2 Risks to the public

Much of the work associated with the street lighting installations will be carried out on existing pedestrian routes which are to remain open at all times. Suitable signs are to be erected warning members of the public of the work being undertaken.

5.8.3 Falling/tripping

All open excavations are to be adequately protected and are to be backfilled as soon as possible. All materials are to be stored away when not required.

5.8.4 General site activities/contractor's staff

Ensure the provision of personal protective equipment.

5.8.5 Manual handling of heavy equipment

Ensure compliance with Manual Handling Operation Regulations 1992.

5.9 Feature screen to plaza housing access features

A 6m high stainless steel feature screen is to be erected in the entrance plaza. The main hazards associated with this aspect of the work include:

5.9.1 Manual handling of heavy equipment

Ensure compliance with Manual Handling Operation Regulations 1992.

5.9.2 On-site welding operations

Ensure safe working methods.

5.10 General manual handling

The contractor's proposals are required for the safe manual handling of such items as:

- railings and gates
- components of octagonal feature piers
- pier caps and wall copings
- lighting columns
- bollards
- inspection chamber covers
- rootballed trees
- seats and other street furniture

5.11 Work to existing trees

The principal contractor is to prepare a method statement for safe working.

5.12 Maintenance of soft landscaped areas

At various stages during the works and following practical completion, the contractor will be required to carry out maintenance work to soft landscaped areas. Problems arising out of such work will include:

- conflict with members of the public;
- the use of vehicles in areas used by the public;
- the use of chemicals in areas used by the public.

Particular care is to be used during such operations and the work is to be carried out by suitably trained and qualified operatives in accordance with all current regulations and enactments.

6. CONSTRUCTION MATERIALS

6.1 Construction materials generally

The following potentially hazardous construction materials and substances are required by the design:

Examples of substances harmful by inhalation:

- welding fumes
- hardwood dust

- cement dust
- isocyanates (paints, varnishes, adhesives)
- solvents (paints, strippers, mastics, glues, surface coatings)

Examples of substances hazardous in contact with skin and mucous membranes:

- bitumen
- brick, concrete, stone dust
- cement
- paints, varnishes, stains
- certain epoxy resins
- chromates (in primer paints, cement)
- petrol, white spirit, thinners
- acids
- alkalis

The principal contractor is required to take appropriate measures to control the risks created by these hazards, to detail these proposed control measures in the health and safety plan and to prepare detailed method statements for managing these aspects of the works.

Other common materials and substances used during construction will also present health and/or safety hazards, requiring the contractor to carry out COSHH or other risk assessments and to introduce control measures. These are deemed to be within the normal experience of a competent contractor and have not therefore been listed here.

6.2 Hot bituminous material

Care is required whilst working with hot bituminous materials. Suitable protective clothing and footwear must be worn.

7. SITE-WIDE ELEMENTS

7.1 Site access and egress points

The main vehicular access to the site will be at the west entrance. A secondary site access is available to the north of the site.

7.2 Movement of site traffic

The principal contractor must liaise with other contractors on adjacent sites regarding the movement of vehicles around the site and around the immediate area.

It will be the responsibility of the principal contractor to detail their proposals for managing the movement of all site traffic, including their own plant, those of subcontractors, delivery vehicles and the like, around the site and around the immediate area. The effects of site traffic on the local environment and upon people living and working in the area are to be kept to a minimum.

These details/method statements will be incorporated into the health and safety plan prior to the work commencing on site.

7.3 Location of temporary site accommodation

The principal contractor's details are required for the proposed location of temporary site accommodation and of the site compound. The proposed location of such accommodation is to be indicated on a drawing which will form part of the construction-phase health and safety plan.

7.4 Location of unloading, layout and storage areas

Vehicles to be loaded/unloaded under supervision by a competent person.

No vehicles will be allowed to wait on main roads.

A method statement is required detailing the procedure for loading/offloading materials and for maintaining stability and tidiness of stacks and containers.

7.5 Traffic/pedestrian routes

All traffic and pedestrian routes, except where closed to the public as indicated on the drawings, must remain open at all times and must remain free from spoil and debris. Vehicle access has to be maintained for the public and for emergency services at all times. All measures are to be taken to minimise conflicts between the public and site traffic.

All routing of site traffic, signing and barriers are to be in accordance with safe site practice and local authority requirements. A stopping-up order will be required for any essential temporary closures.

The contractor is to prepare method statements for phasing and safe working.

7.6 Site security

The site is currently secured by a 2.4m high hoarding around the perimeter of the site. The contractor is to allow for regular inspection and maintenance of all such hoardings during the works.

At various stages of the works several lengths of the security hoardings will need to be dismantled and re-erected to allow certain aspects of the work to be carried out. The permanent removal of this hoarding will form part of these works; they are to be dismantled:

(a) on completion of contract works, or
(b) on completion of permanent boundaries, or
(c) to facilitate works, in which event alternative measures are to be taken to ensure public safety.

7.7 Co-ordination with work to be executed by others

The following work is to be carried out by other contractors:

7.7.1 Installation of play equipment

7.7.2 Outfall to housing development

The contractor is to obtain from, and arrange with, the other contractors details of times of commencement of work and delivery of plant and materials, phased to suit the contractor's programme and must allow all reasonable access.

The contractor is to be the principal contractor for these aspects of the work. Safe methods of work are to be agreed between all contractors. All personnel are to be made aware of general health and safety requirements demanded elsewhere in the health and safety plan.

7.8 Site welding

Site welding operations will be required to the components of the feature screen to the plaza and to other aspects of the works. All precautions are to be taken to prevent injury not only to the personnel carrying out the works but also to personnel not actually engaged in the operation and also to members of the public.

8. OVERLAP WITH CLIENT'S UNDERTAKING

8.1 Buildings in occupation

It must be acknowledged and accepted by the principal contractor that adjacent properties will remain occupied throughout the duration of the contract. Specific provision must be made for the protection of the public, especially while contractor's plant is moving around the area.

All existing services must be maintained.

The contractor's proposals are required for managing these problems.

9. SITE RULES

9.1 Generally

The principal contractor is reminded of the high public profile which the client enjoys and shall therefore ensure that no site activity reflects adversely on the client.

The principal contractor's proposals for complying with these site rules are to be set out in the construction-phase health and safety plan and must include provisions for:

9.2 Conformity with all statutory requirements

Including:

- The Construction (Health, Safety and Welfare) Regulations 1996
- Electricity at Work Regulations
- PPE at Work Regulations
- COSHH Regulations
- Noise at Work Regulations

9.3 Booking in/out

The contractor's proposals are required for ascertaining what visitors are present on site and ensuring they are aware of the site risks.

9.4 Subsoil stockpile

The provision of a subsoil stockpile may mean double handling of material, which increases the possibility of site traffic conflicts and may also impose restrictions on site working.

Work is to be carried out by suitably trained and qualified operatives. The contractor is to allow for all effects in programming works and is to prepare a method statement for safe working, identifying any temporary stockpile areas required.

9.5 Topsoil storage area

The risks involved include site traffic conflicts and conflict with members of the public.

Work is to be carried out by suitably trained and qualified operatives.

The contractor is to prepare a method statement for safe working, identifying proposed phasing of the works and haulage routes.

9.6 Control of movements

The contractor's proposals are required for controlling the movements around the site of all occasional personnel.

9.7 'Absence' procedure

The contractor's proposals are required for appointing a person responsible for ensuring the management of health and safety on site in the absence of the named person in the health and safety plan.

The proposals are to cover short absences of less than one day and for planned and unplanned absences for one full day or more.

9.8 Unforeseen eventualities

A formal reporting system for unforeseen eventualities is required.

9.9 Fires on site

Fires will not be permitted on the site.

9.10 Site security

The contractor's proposals are required for ensuring that the public and the like are kept off site, especially outside normal working hours. Detailed requirements are set out in the bills of quantities. The contractor is to state how this aspect of security is to be managed.

The contractor is required to detail proposals for dealing with regular inspections and repair of damaged or vandalised fencing.

9.11 Access around the site

The contractor's proposals are required for identifying walkways and routes and ensuring that they remain clear and unrestricted at all times. Proposals should cover all operatives and visitors for all stages of construction.

The contractor's proposals are required for clearing away all debris and leaving the site reasonably tidy at the end of each working day. Rubbish and scattered building materials will not be allowed to accumulate.

9.12 Wheel wash facility

The contractor is to set up a wheel wash facility to ensure that all roads and footpaths remain free from spoil and debris at all times.

9.13 Provision of welfare facilities

Provide all necessary welfare facilities including sanitary conveniences and washing facilities to comply with Regulation 22 of the Construction (Health, Safety and Welfare) Regulations.

9.14 First aid

Adequate first aid equipment is to be made available and the contractor must ensure the presence on site of a person trained in its use at all times.

Specific proposals are required for the treatment of the following:

(a) the risk of drowning and associated hazards;

(b) the risk of electrocution especially where electrically operated tools are used in the vicinity of water; and

(c) the recognition and preliminary treatment of leptospirosis.

9.15 Personnel on site

Considering the remoteness of certain areas of the site, no personnel are to be allowed to work alone in such locations. This particularly applies where work is being undertaken in close proximity to the watercourses.

9.16 Communication around the site

The use of mobile phones and/or walkie-talkies is recommended for communication between remote areas and for use in emergencies.

9.17 Control of noise

The contractor's proposals are required for ensuring that limits described in the Preliminaries are not exceeded.

The contractor's proposals are also required for ascertaining noise levels on site at regular intervals and when requested by the client's representative.

Comply generally with BS5228 and all relevant legislation.

Fit all compressors, percussion tools and vehicles with effective silencers of a type recommended by manufacturers of the equipment.

Other possible hazards to be taken into consideration include vibrations and the effect that these have on the workforce as well as on the public. Ear protection should be considered during high level noise operations.

9.18 Control of pollution

The contractor's proposals are required for ensuring that the site is kept clean and for ensuring the control of dust.

The contractor's proposals are required for ensuring that debris does not fall into the watercourses.

9.19 Training

The contractor's proposals are required for induction, ascertaining general needs and specific training for identified risks.

9.20 Competence of site personnel

The contractor's proposals are required for vetting and for recording the competence of all personnel and other contractors' personnel for all stages of construction and management of the site.

10. CONTINUING LIAISON

10.1 Generally

Procedures for dealing with unforeseen events during the project which result in substantial design changes and which might affect resources are as follows:

10.1.1 In the event of any unforeseen circumstance, the planning supervisor is to be informed immediately by the principal contractor.

> **10.1.2** Details of the health and safety issues arising from any unforeseen occur-
> rence are to be submitted to the planning supervisor as soon as is
> possible after the event.
>
> **10.1.3** In the event that any redesign is required, for whatever reason, the
> health and safety implications are to be submitted for consideration and
> acceptance by the planning supervisor in due time before execution.
>
> ## 10.2 Recording health and safety matters
>
> The principal contractor's proposals are required for recording and resolving
> health and safety matters realised by contractor's personnel, other contractors,
> client's representatives, members of the public and any other affected persons.
>
> ## 11. TENDER STAGE METHOD STATEMENTS
>
> Tender stage method statements must be submitted with the tender *outlining* safe
> working methods for the following:
>
> - Excavation below and beyond the remediation level;
> - Work in proximity to the watercourses;
> - Work on or adjacent to public rights of way;
> - Co-ordination with work to be executed by other contractors on adjacent
>   developments;
> - Haulage and site traffic routes;
> - Stockpile areas;
> - Co-ordination with work to be executed by others (such as play equipment,
>   outfall from housing development).
>
> The method statements must describe safe working methods for these items and
> must also describe how and when the contractor proposes to undertake the work.
>
> The principal contractor may, at his discretion and at the same time, submit method
> statements for other parts of the works.

Part E: Contractor's Construction-Phase Health & Safety Plan

CONSTRUCTION-PHASE HEALTH AND SAFETY PLAN

INTRODUCTION

This health and safety plan has been produced to detail the contractor's approach to the control of safety on the project. The layout of the plan is in accordance with the requirements of the Construction (Design and Management) Regulations 1994.

This health and safety plan has been developed from the pre-tender health and safety plan prepared by the planning supervisor.

Relevant Legislation:

- Workplace (Health, Safety and Welfare) Regulations 1992
- Management of Health & Safety at Work Regulations 1992, amended 1994
- Personal Protective Equipment at Work Regulations 1992
- Manual Handling Operations Regulations 1992
- Provision and Use of Work Equipment Regulations 1992
- Health & Safety (Display Screen Equipment) Regulations 1992
- Control of Substances Hazardous to Health Regulations 1988, amended 1991
- Health & Safety (First Aid) Regulations 1981
- Reporting of Injuries, Diseases and Dangerous Occurrences Regulations 1995
- Health & Safety (Information to Employees) Regulations 1989
- Road Traffic (Carriage of Dangerous Substances in Packages etc.) Regulations 1992
- Road Traffic (Training of Drivers of Vehicles Carrying Dangerous Goods) Regulations 1992
- Noise at Work Regulations 1989
- Construction (Lifting Operations) Regulations 1961
- Construction (Head Protection) Regulations 1989
- Construction (Design and Management) Regulations 1994
- Construction (Health, Safety and Welfare) Regulations 1996
- Environmental Protection Act 1990
- Waste Management Licensing Regulations 1994, amended 1995
- Control of Pollution (Amendment) Act 1989
- Controlled Waste (Registration of Carriers and Seizure of Vehicles) Regulations 1991
- Water Resources Act 1991
- Water Industry Act 1991

PROJECT POLICY

It is the policy of the organisation that we will at all times conduct our activities so that we will, as far as possible, carry out the works so that they exceed the minimum requirements of the relevant legislation.

We will ensure that optimum priority is given to the health & safety of employees and all other persons affected directly by our activities.

We will ensure that every effort is made to minimise any negative impact upon the environment which any activities may have.

Every employee of the organisation has direct responsibility for the health & safety of all persons involved by our actions.

Good safety practices will be a primary concern on this project.

DESCRIPTION OF WORKS

Civil engineering, external and landscape works including earthmoving and fencing to approaches and periphery of the development.

These include:

- earthmoving and topsoiling
- planting trees and shrubs
- grass seeding
- fencing works
- maintenance works

PROJECT INFORMATION

Project Name:

Site Address:

Contact:
Site Telephone No:

Client:

Contact:
Telephone:
Fax:

Planning Supervisor:

Contact:
Telephone:
Fax:

Principal Contractor:

Contact:
Telephone:
Fax:

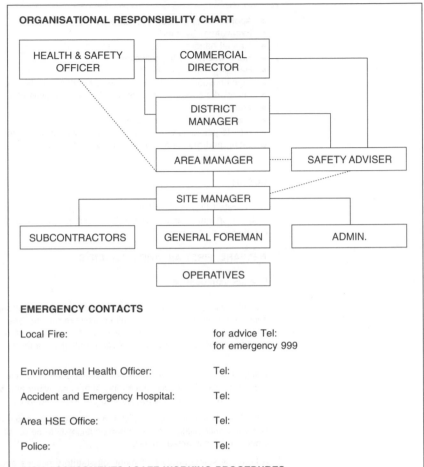

ORGANISATIONAL RESPONSIBILITY CHART

EMERGENCY CONTACTS

Local Fire: for advice Tel:
                                               for emergency 999

Environmental Health Officer: Tel:

Accident and Emergency Hospital: Tel:

Area HSE Office: Tel:

Police: Tel:

RISK ASSESSMENTS / SAFE WORKING PROCEDURES

HAZARD IDENTIFICATION

Generic hazards identified for which standard risk assessment records have been prepared and are provided for all personnel on site.

- Unloading/loading by machine
- Grass cutting
- Use of tractors and attachments
- Loading and securing power machinery on to transport
- Pedestrian-operated power equipment
- Excavation by machinery
- Transport of materials and equipment
- Application of chemicals
- Use of non-powered hand tools
- Use of powered hand tools
- Working above ground level
- Carrying out operations near roads or water bodies
- Maintenance of vehicles and machinery
- Fence wire tensioning

- Application of protective treatments
- Excavations by hand
- Manual handling
- Hand weeding, litter collection
- Handling cement by hand
- High/unusual risk activities arising from this specific site
- Site/works specific (identified by planning supervisor)
- Working on slopes
- Lighting and cable installation
- Risk to pedestrians in work associated with pavements
- Lifting and manoeuvring heavy/awkward objects such as pergola, inscription panel and rootballed trees.

COSHH

A separate COSHH file will be kept on site for all materials/products incorporated and/or used on the contract. This will be incorporated into the Health & Safety File on completion of the works.

WELFARE, FIRST AID AND ACCIDENTS

Site accommodation

The site will be kept secure for the storage of equipment and for the provision of appropriate welfare facilities for all employees on site. Secure storage will be provided for work clothing and PPE for all those for whom it is required. Toilet facilities are provided within the compound, adequate for the numbers of personnel likely to be on site at any one time.

Within the compound there will be accommodation for the taking of meals which are brought to site. There will be a supply of drinking water and sufficient water provided for the maximum numbers of people who are expected to be on site.

A qualified first aider is employed and will be based on site with an appropriate first aid kit for the anticipated number of personnel likely to be on site at any one time. The first aider can be contacted on Tel:

A health & safety policy statement, all statutory notices and employees' liability insurance certificate will be displayed within the accommodation. Abstracts from the relevant legislation may, if appropriate, also be displayed.

ACCIDENT REPORTING PROCEDURES

An Accident Book will be kept on site at all times. The procedure followed will be as described in the attached 'INCIDENT REPORTING PROCEDURE'.

MANAGEMENT OF SUBCONTRACTORS

All subcontractors are to complete the pre-qualification questionnaire for contractors to assess suitability and competence.

Prior to work starting on site a pre-contract meeting is to be held covering the following topics:

- Induction of employees
- Risk assessments and safe working procedures
- Control of hazardous substances
- Personal protective clothing/equipment
- Operations and use of plant and machinery
- Welfare and first aid arrangements

Site induction

All operatives and management will undergo induction training prior to being allowed to work on site. A standard checklist will be used for:

- site details, i.e. address, telephone and fax numbers
- safety responsibilities to safety management team
- site security arrangements and working hours
- location of first aid equipment in site office
- name of trained first aider
- canteen/messroom arrangements
- toilet and washing facilities
- smoking policy
- fire procedures
  - escape routes
  - exits
  - assembly points
  - location and type of fire extinguishers
- disciplinary procedures
- risk assessment/safe working procedures applicable to employees
- site safety rules

Mechanical plant

Mechanical plant will be used only by operatives trained in its use. Hired plant must:

- arrive on site with confirmation that it has been maintained to the required standard;
- must be maintained in good working order and free from defects while on site.

Operators with hired plant must be competent and able to demonstrate their training background with recognised certification.

Copies of all certificates will be kept in the site office. Equipment requiring certification will not be permitted on site unless it is accompanied by a current certificate.

SITE TRAFFIC / SITE RESTRICTIONS

A number of roads around the site are available to other users and therefore, in order to reduce the risk of injury, certain precautions will be observed as follows:

- All site personnel and visitors will wear high-visibility vests on site at all times.
- All lorries will be controlled by a banksman while manoeuvring on site.
- An on-site speed limit of 15 mph will be instructed and enforced.
- All drivers of site vehicles or plant will be in possession of a current full driving licence.

EXCAVATION

It is not anticipated that any excavations in excess of 2.4m depth will be required during the course of this project. If any are required then appropriate precautions will be taken and further risk assessments and safe working procedures prepared.

It is envisaged that all excavation below levels of remediation will involve the use of machinery. The safe working procedure applicable will apply with the following additional precautions:

- All operators, banksman and workmen involved in excavation work to be briefed on the implications as described in the planning supervisor's report.
- All substances other than standard stone/soil fill material are to be reported to the site supervisor immediately they are revealed. Excavation should cease until further notice.

- Special attention is to be given to all liquids, metals, containers, coloured glazing of the soil and odours emanating from the excavation.

EXISTING SERVICES

Where there is a likelihood of encountering any services the appropriate action will be taken:

Hazardous activities are as follows:

- Excavation by machine
- Excavation by hand
- Installation of electrical services

Precautions necessary:

- See safe working procedures
- Electrical contractor to prepare risk assessments and safe working procedures
- Prior to excavation all services marked out
- Trial holes to be dug prior to deep excavations in the vicinity of the services
- Cable detection equipment and banksman to be used in all circumstances

CONFINED SPACES

Any operation which requires entry into a confined space will require a PERMIT TO WORK.

Prior to the work being carried out, a comprehensive hazard identification and risk assessment will be carried out in consultation with our safety adviser.

Any necessary escape equipment which is identified as being required will be obtained prior to the work being commenced.

The permit to work will be issued by the project manager and will be kept in the site office.

During the operation of a permit to work a sign will be displayed which forbids entry to unauthorised personnel.

NOISE

It is not anticipated that noise levels will exceed unobtrusive levels, but random assessments will be carried out and if any problem is identified action will be taken as necessary.

MONITORING PROCEDURES

The site manager's responsibilities include inspection to ensure that health & safety standards on site are maintained. A site safety inspection report form is to be completed at least weekly. Enforcement notices can be served by the site manager either:

- an immediate action notice requiring the contractor to confirm the remedial action taken; or
- a stop notice requiring the contractor to cease the activity for repeated contravention or for a serious breach of health & safety.

SAFETY PLAN REVIEW

This safety plan will be reviewed regularly to ensure that the information is up to date.

Items of on-going management are identification of further hazards, necessity for additional procedures and alterations to training needs.

References

British Standards Institution (BSI) (1997) *BS8750: Guide to Occupational Health and Safety Management Systems*, BSI.

Croner (1994) *Croner's Management of Construction Safety*, Croner Publications.

European Commission (1989) Directive 89/391/EEC on the introduction of measures to encourage improvements in the health and safety of workers at work, HMSO.

Health & Safety Executive (HSE) (1961a) *The Factories Act 1961*, HMSO.

Health & Safety Executive (HSE) (1961b) *The Construction (General Provisions) Regulations 1961*, HMSO.

Health & Safety Executive (HSE) (1961c) *The Construction (Lifting Operations) Regulations 1961*, HMSO.

Health & Safety Executive (HSE) (1966a) *The Construction (Working Places) Regulations 1966*, HMSO.

Health & Safety Executive (HSE) (1966b) *The Construction (Health and Welfare) Regulations 1966*, HMSO.

Health & Safety Executive (HSE) (1974) *The Health and Safety at Work, etc. Act 1974*, HMSO.

Health & Safety Executive (HSE) (1988) *Blackspot Construction*, HMSO.

Health and Safety Executive (HSE) (1992a) *The Management of Health and Safety at Work Regulations 1992*, HMSO.

Health and Safety Executive (HSE) (1992b) *The Workplace (Health, Safety and Welfare) Regulations 1992*, HMSO.

Health and Safety Executive (HSE) (1992c) *The Provision and Use of Work Equipment Regulations 1992*, HMSO.

Health and Safety Executive (HSE) (1992d) *The Personal Protective Equipment at Work Regulations 1992*, HMSO.

Health and Safety Executive (HSE) (1992e) *The Manual Handling Operations Regulations 1992*, HMSO.

Health and Safety Executive (HSE) (1992f) *The Health and Safety (Display Screen Equipment) Regulations 1992*, HMSO.

Health & Safety Executive (HSE) (1994a) *CDM Regulations: How the Regulations Affect You*, HMSO.

Health & Safety Executive (HSE) (1994b) *The Construction (Design and Management) Regulations 1994*, HMSO.

Health and Safety Executive (HSE) (1995a) *The Reporting of Injuries, Diseases and Dangerous Occurrences Regulations 1995*, HMSO.

Health and Safety Executive (HSE) (1995b) *A Guide to Managing Health and Safety in Construction*, HMSO.

Health and Safety Executive (HSE) (1995c) *Managing Construction for Health and Safety*, HMSO.

Health & Safety Commission (HSC) (1996) *Health and Safety Statistics 1995/96*, HMSO.

Royal Institute of British Architects (RIBA) (1992) *Standard Form of Agreement for the Appointment of an Architect (SFA/92)*, RIBA.

Royal Institute of British Architects (RIBA) (1995a) *Form of Appointments as Planning Supervisor (PS/45)*, RIBA.

Royal Institute of British Architects (RIBA) (1995b) *Conditions of Engagement for the Appointment of an Architect (CE/95)*, RIBA.

6 Environmental evaluation and control systems

Introduction

There can be little doubt that almost all organisations in the construction industry, like virtually all businesses in the industrial, manufacturing and commercial sectors, face increasing pressures to expand their environmental awareness, improve their environmental performance and provide tangible measures of environmental safeguarding. Such pressures come from clients, investors, consumers, regulatory bodies and, more generally, the public. The construction industry has an unequivocal and predominant effect upon the environment whenever and wherever construction works are carried out. Spiralling demands for more environmentally empathetic and sustainable construction and, in particular, more stringent regulation mean that *environmental management* will become a prominent consideration in the management of construction processes in the future. This interest may follow in much the same way that the management of quality has become an established aspect of construction management over the last 20 years.

Awareness for environmental management within the construction industry is increasing. This was stimulated by the introduction, in 1992, of BS 7750: *Specification for Environmental Management Systems* and its counterpart ISO 14001. Environmental management systems, commonly referred to as 'EMS', are beginning to be considered by construction clients as an important aspect of project evaluation and development. The concepts are also being adopted by companies in the service and supply sectors.

The adoption of environmental management in construction has been somewhat slow relative to its adoption in other industrial sectors. This is not unexpected, given that a widespread recognition of environmental management can only follow a culture change similar to that which accompanied the evolution of quality management during the 1980s and 1990s. It is appreciated that some of the demoralising early experiences with quality systems development and implementation may have fostered a lack of initial interest in environmental management practices. Nevertheless, some organisations have undergone a culture change and begun to address the likely future demands for environmental management. Their experiences have been positive, and it is such organisations which will be best placed to meet the environment-related business challenges of the future.

An organisation does not have to introduce an all-embracing and complex environmental management system in one go: it can build up a system gradually. An incremental approach can be less divisive to the organisation while facilitating the desired and necessary culture change.

An environmental management system can be developed by bringing existing policies and procedures under a focused environmental umbrella. Alternatively, new working practices and managerial approaches may need to be developed. Systems can include 'standard' systems – those meeting the BS 7750/ISO 14001 specification or the European Eco-Management and Audit Scheme (EMAS) (DoE, 1995); and 'bespoke' systems – those tailored specifically to the needs of the organisation.

The contribution made to an organisation through the establishment of an environmental management system can be significant. Benefits can extend well beyond merely satisfying environmental regulations. There are real benefits to be gained by an organisation in both its internal operations and its external business. Examples include better corporate focus, efficiency improvements, enhanced credibility, reduced liability and competitive advantage.

Application of environmental management to construction projects is increasing. Since the mid-1980s, *environmental impact assessment* (EIA), 'a systematic process that examines the environmental consequences of development actions, in advance' (Glasson *et al.*, 1994), has been applied to the project evaluation stage of construction through the physical planning processes. More recently, environmental management has become a part of the procurement strategies of some clients (CIRIA, 1995). One such example is that of a government client who, when pre-qualifying prospective lead design consultants and principal contractors for a major project, requested a formalised environmental management system encompassing design, construction and contract administration which could be extended to operation and maintenance; an environmental plan; a policy on environmental resourcing; and a set of construction environmental practices. In the broadest sense, environmental management is a management process that extends across the total construction process, from project evaluation, through design, procurement and contract administration, into use and upkeep of the building or structure.

It is likely that the main pressures on construction consultants and contractors to support environmental management will come from public sector and major private sector clients procuring large projects, in particular those projects with significant environmental sensitivity. In time, it is possible that having an environmental management system will be the deciding factor between organisations being considered for such projects. Such systems will increasingly be established by construction companies to avoid competitive disadvantage. In addition, the introduction of more rigorous environmental legislation is likely to increase the popularity of environmental management systems as they can represent an effective and recognised response.

However, the construction industry is behind other industry sectors in the adoption of environmental management concepts. A likely reason for this is that, currently, organisations do not or cannot perceive the cost-effectiveness of environmental management systems, which provide as many soft benefits as they do hard benefits.

In addition, the demands of environmental management require significant changes to organisational awareness, understanding, effort and commitment. All this comes at a time when organisations are already consumed in complying with very demanding health and safety legislation and meeting escalating client demands for better project quality, efficiency and cost-effectiveness. Nevertheless, environmental management, through the establishment of environmental management systems, will in the future be one of the most challenging aspects for the construction industry to positively embrace.

This chapter focuses on environmental evaluation and control. In synopsis, aspects explained are

- the relevance of environmental management to construction;
- the need for environmental management systems;
- environmental management standards;
- environmental management and the corporate organisation;
- environmental management and the project organisation;
- environmental evaluation during project evaluation and development; and
- environmental control during construction by the contracting organisation.

Environmental management and construction

The environmental effects of construction

Organisations interact with their environment in many ways and to varying degrees. The effects that an organisation can have upon its environs can be direct and indirect, beneficial or adverse, and ambiguous or manifest. All organisations will, at some time, give rise to environmental effects as a result of their business activities. Therefore, environmental management should be a fundamental consideration for every organisation. Perhaps more so than any other industry, construction has an undisputed and considerable effect upon the environment.

The effects of the construction process are great. There are always extensive environmental effects at the project workplace. These are the direct effects that everyone can see: for example, land use, deforestation, pollution and noise. There are also the indirect effects on the environment by the production processes involved in those materials and products that the construction industry consumes: for example, the use of natural resources, many of which are non-renewable.

The environmental effects of construction do not end with the completion of the construction phase. A building or structure has a lasting effect on the populace and environment in which it is placed. Furthermore, the use of the product has tremendous long-term effects on its surroundings. Environmental effects are, therefore, synonymous with the construction industry.

Construction has an intrinsic and enormous propensity to affect the environment in both useful and beneficial ways, but more often than not in unnecessary and detrimental ways. Environmental management is an essential tool in ensuring that the positive aspects are enhanced and negative facets minimised. To be effective in the long term, the concepts must become a prerequisite to the future structure, organisation and management of construction. This is essential if both the major and instantly recognisable environmental effects and those effects that are less easy to see are to be better managed in the future.

The most prominent environmental effects of construction can be grouped as shown in Table 6.1.

Environmental awareness

Recent change in business thinking and activity has, in general, followed a profound social change in attitudes. People have begun to appreciate more fully and want to safeguard the environment in which they live. A greater awareness of environmental issues, environmental standards, environmental management and environmental management systems by organisations is essential in managing the effects of construction upon the environment and population more effectively.

This is further emphasised by the following:

- The public is becoming increasingly intolerant of environmentally unsound business practices by organisations they perceive to be threatening the environment.
- Growth in the environmental or 'green' movement at national and international levels is putting pressure on governments and industries to be more environmentally responsible and become more accountable for their actions.
- National and international environmental legislation affecting organisations in all business sectors is becoming increasingly stringent.
- Environmental issues are becoming increasingly important in the commercial image and competitive marketing of organisations.

In response to the challenge invited by greater environmental awareness, the construction industry must look towards improving its products and services, procedures and practices across the total construction process. It must do this by focusing on better 'eco-construction'. This is concerned with devising new methods or adapting existing methods of undertaking the total construction process with greater empathy for the environment. This means that project evaluation, design, construction and use of the finished construction product should be conducted within a framework of environmental management.

Table 6.1 The environmental effects of construction

Environmental effect	Implications
Land use	All construction projects invariably occupy land, consume space above land and use areas below land. This is, without doubt, the most prominent environmental effect of construction.
Existing site dereliction	The preference to develop green-field sites leaves many localities with derelict sites and redundant buildings and structures in ruinous condition.
Natural habitat destruction	Many construction projects leave their mark on the landscape, natural amenities and wildlife, much of which once destroyed can never be replaced.
Use of natural resources	The use and destruction of natural resources is a severely detrimental by-product of construction. For example, deforestation and quarrying leave their mark on the environment long after the raw materials have been extracted and used in the construction process.
Air emissions and pollution	The construction process on site gives rise to many atmospheric pollutants in the form of toxic fumes from plant and equipment use and the production processes employed. In addition, in a completed building for example, there can be many sources of air emission such as chlorofluorocarbons (CFCs).
Use of water resources	As a by-product of all new development, increasing demands are placed upon water supplies. This may be over-use of existing supply points or the requirement to build further supply facilities, which themselves commence a cycle of environmental effects.
Discharges and water pollution	Site production processes frequently give rise to the spillage and discharge of contaminants through natural water courses and man-made drainage systems.
Waste	Waste as a result of acquiring raw materials, refining raw materials for use in the construction process, delivering material to site, and poor storage, handling and use of materials during the construction phase all give rise to the large amount of waste created within construction.
Comfort disturbance	Noise, dust, dirt, pollution and traffic are some of the most prominent comfort disturbances experienced by local inhabitants to any construction project, and despite the usual measures taken by the contractor they remain intrinsic to the site processes.
Health and safety	By its very nature construction is accompanied by an inherent level of danger to both the employees working on the site and members of the local public community.
Energy consumption	Construction gives rise to a considerable level of energy consumption both during the construction processes and in the use of the finished product.

Furthermore, it implies that each contributor of professional services or products should support environmental management within their organisation to ensure their environmentally sympathetic contribution to the construction process. Although such a view might be considered to be somewhat idealistic, it should be remembered that the concept of quality management systems (QMS) was at a similar stage in the early 1980s and that the quality movement has since developed to the point where quality management has become a necessary and accepted part of the construction process.

Environmental consideration

Respect for and necessary consideration given to the environment influences construction in many diverse and significant ways (Griffith, 1995). Considerations can be grouped into four broad areas:

1. *The environment is directly and greatly affected by all construction projects at their site.* Because this is so, the potential for environmental effects must be considered before any approval is given for new development. Environmental impact assessment (EIA) exists for this purpose. This aspect of environmental management is the responsibility of the developer or client and is actioned through the role of the lead consultant in association with specialist environmental consultants.

2. *The environment is affected directly and indirectly by construction materials and products.* Because this is so, environmental management must be a part of the process of manufacture and provision of construction materials and components. A limited but growing number of suppliers support the concept of green or 'eco-labelling' – provision of products that are advantageous to environmental protection. These organisations apply EMS concepts in the production and supply processes.

3. *The environment is affected directly and greatly by the processes and management of construction.* Because this is so, environmental management must be a part of contract administration on site. Contracting organisations which operate an EMS will not only understand better the need for environmental management in construction but also take advantage of the benefits that it can bring. Moreover, they will also be able to translate environmental management concepts from the corporate perspective into the contract administration phase of their projects.

4. *The environment is affected directly and often greatly through the use of the completed construction project.* Because this is so, environmental management should be a part of the maintenance and facilities management of the development. Occupiers or users of the development, who are frequently the client, and who operate an EMS within their organisation, are likely to achieve better utilisation and maintenance of their capital asset resources.

Combating environmental effects through environmental management systems

The promotion and success of formalised environmental management in combating the environmental effects of the construction industry will be determined by two principal influences:

- *Environmental legislation* – increasingly stringent environmental legislation will place a greater responsibility on all contractual parties within the construction process to improve their environmental performance.
- *Environmental commitment* – all organisations must want to be proactive in environmental protection and seek to improve their environmental performance through the establishment of an appropriate management system.

Environmental management will address the environmental effects of construction at the regulatory, organisational and project levels (see Figure 6.1). Three aspects are therefore of particular importance:

1. *Environmental impact assessment (EIA)* – the environmental evaluation, undertaken by a developer or client during project evaluation and development, of the potential effects of a proposed project.
2. *Environmental management system (EMS)* – the contribution by each participant to the total construction process through the use of a formalised management system.
3. *Environmental management programming* – the detailed evaluation of the project by the contractor to identify the optimum environmental approach and the determination of requirements, responsibilities and mechanisms to ensure environmental control during the works.

These aspects, although quite separate in terms of their undertaking because they involve different contractual participants, are quite closely linked. For environmental management to be effective throughout the total construction process, each of the aspects must be given careful consideration at the appropriate point in the process.

The principal issue is that unless all participants in the construction process actively seek to improve their individual environmental performance and in so doing contribute to the environmental performance of the construction project the impetus for environmental management will effectively be lost.

The need for environmental management and EMS

The need for environmental management exists in any construction organisation that seeks to:

- recognise the growing concern for environmental issues;
- determine the environmental effects of its business;
- demonstrate its commitment to environmental performance;
- safeguard the environment;
- comply with its environmental policy, strategy, aims and objectives;

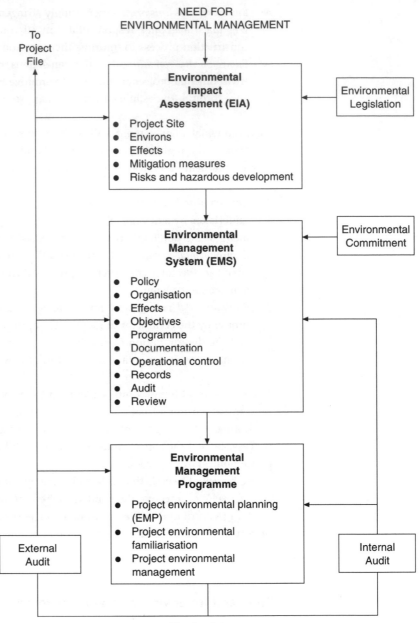

NEED FOR
ENVIRONMENTAL MANAGEMENT

To
Project
File

**Environmental
Impact
Assessment (EIA)**

- Project Site
- Environs
- Effects
- Mitigation measures
- Risks and hazardous development

Environmental
Legislation

**Environmental
Management
System (EMS)**

- Policy
- Organisation
- Effects
- Objectives
- Programme
- Documentation
- Operational control
- Records
- Audit
- Review

Environmental
Commitment

**Environmental
Management
Programme**

- Project environmental planning
  (EMP)
- Project environmental
  familiarisation
- Project environmental
  management

External
Audit

Internal
Audit

Fig. 6.1 Framework for
project environmental
management

Feedback loops

- accommodate increasingly stringent environmental legislation;
- demonstrate its environmental position to a national and international audience;
- be prepared for environmental assessment, pre-qualification and auditing of its business;
- achieve recognition for its environmental initiatives.

As all organisations are likely in the future to respond, through business necessity, to the strongly developing concern for better environmental performance and a greater level of environmental protection, so support for environmental management will gain greater momentum. Organisations may find that proactive participation in environmental management will be a prerequisite to their level of competitiveness or even their business survival.

It is clear that in some European countries, North America and Japan that the early establishment of environmental legislation has given industries there a commercial advantage in worldwide markets. Companies in those countries have a headstart in environmental management as legislation begins to tighten. As the construction industry will, without doubt, face tougher environmental standards, it will be those organisations which support environmental management and use EMS which will respond best to the demands and which are most likely to be successful.

Environmental management systems and standards

Environmental management systems

An environmental management system (EMS) is 'the organisation's formal structure that implements environmental management' (Griffith, 1997).

BS 7750/ISO 14001: *Specification for Environmental Management Systems* requires that the organisation develop, implement and maintain an EMS to ensure that its activities conform to the environmental policy, strategy, aims and objectives that it has set. Moreover, it should warrant that the system meets all current environmental legislation that regulates its business activities. The organisational ethos should be founded on the premise that all businesses impact upon their environment; management systems control all aspects of business; and standards can be set for and achieved by environmental management systems. Such an ethos must be championed by commitment from the top of any organisation and become the shared goal of the organisation.

To establish an EMS, an organisation should seek to:

- Develop an 'environmental management systems manual' – a documented set of organisational policies, procedures and working instructions that satisfy the environmental standard.
- Implement the requisite procedures and working instructions in the course of its on-going business operations and educate the organisation in their effective use.
- Maintain the procedures and working instructions and audit, review and upgrade them on a continuous basis to ensure that they continue to meet policies and standards.

The achievement of these requirements is fundamental to the internal development and implementation of the EMS by the organisation. Furthermore, it is essential in demonstrating to external parties that the organisation is up-holding its commitment to environmental issues. This is important when the organisation is seeking wider recognition for its environmental initiatives: for example, when pre-qualification is demanded by clients and, in the future, when it may pursue certification for its EMS.

An EMS can, if so desired, be quite complex in nature. Its level of complexity is principally a function of the complexity of the organisation itself. The system must consider not only the environmental aspects, which are themselves influenced by a multitude of variables, but also the technological and human characteristics of the organisation. Establishing an EMS therefore requires a comprehensive framework of structure and procedures and also an appreciation of the range of available skills, abilities and commitment within the organisation. However, organisations do not have to develop and implement an extensive and complex system. It may, indeed, be far more practical for an organisation to build up the EMS in steps as organisational needs evolve. In addition, a bespoke system tailored specifically to the needs of the organisation may be preferred to a standard system. While some organisations will be developing an EMS from the baseline, other organisations may already be using management systems that can be used as an administrative umbrella or be modified specifically as a template for EMS application. BS 7750/ISO 14001 suggests the sharing of organisational systems and resources, provided the existing system is itself recognised, for example an ISO 9000 QMS.

BS 7750/ISO 14001 sets out and describes the basic requirements for an EMS in key sections. In these sections, each requirement is framed in terms of organisational responsibility, around which the organisation can develop the system to suit its own business activities. In this way, the individuality and core business of the organisation can be satisfied. This means that the EMS is unique to the organisation. As this is so, the organisation may need to develop some aspects of the system to a greater extent than the minimum requirements specified in the standard. In addition to following the standard, an organisation may need to refer to other authoritative guidance documentation in establishing its EMS. Examples of such documentation include specialist industry sector requirements and the specifications set by particular regulatory bodies.

For an organisation to maximise the use of an EMS, the concepts of environmental management and its application within the organisation should be viewed from a broad perspective. At a primary level, an EMS may encourage improvements within the organisation, for example by suggesting energy saving within the work environment. At a secondary level, an EMS may highlight procedures and practices which might be improved to, for example, reduce waste or encourage recycling of materials and resources. At a tertiary level, an EMS can be used as a marketing and public relations tool to promote the organisation's image in the commercial marketplace. Therefore, to maximise the potential of an EMS the organisation should consider both the internal and external aspects in its development and implementation.

Environmental management standards

BS 7750/ISO 14001 presents a specification for and guidance to any organisation in any sector of business seeking to establish an EMS. It is designed to enable an organisation to develop, implement and maintain an appropriate management system as a basis for ensuring the effective environmental performance of its business practices. The framework of the standards draw upon ISO 9000: *Specification for Quality Systems*. The standards share common systems management principles. It is because of this that an organisation can develop its EMS from an existing and recognised management system template.

The standard presents the key elements that should form part of any EMS. The extent to which any particular element needs to be developed within a system must be determined by the organisation itself. This will depend upon the nature of the organisation and its particular environmental policy, strategy, aims and objectives.

The key elements of an EMS as specified and enumerated in BS 7750/ISO 14001 are as follows:

4.1   Environmental management system
4.2   Environmental policy
4.3   Organisation and personnel
4.4   Environmental effects
4.5   Environmental objectives and targets
4.6   Environmental management programme
4.7   Environmental management manual and documentation
4.8   Operational control
4.9   Environmental management records
4.10 Environmental management audits
4.11 Environmental management reviews

In addition to specifying the key elements of an EMS, the standard presents three annexes:

Annex A   Guide to Environmental Management System Requirements – a general guide to establishing an EMS within an organisation.

Annex B   Links to BS 5750/ISO 9000: Quality Systems – to assist organisations operating a QMS to that standard and who seek to extend that system in accordance with BS 7750/ISO 14001.

Annex C   Links to the European Eco-Management and Audit Regulations – to assist organisations seeking to develop an EMS with a view to registering the system with the European Eco-Management and Audit Scheme (EMAS).

Environmental certification

A company may feel that environmental management is predominantly a function of and benefits its internal organisation. Alternatively, a company may seek to exploit environmental management as an external marketing and public relations opportunity. The type of EMS, the detail of its nature and the level of recognition that the organisation seeks for its EMS is influenced by the organisation's perceived business needs and the likely benefits it can enjoy in the marketplace.

A small organisation, which perhaps operates in a restricted market where formalised environmental management is not a requirement for securing its work base, may have no need to establish an EMS. Conversely, a large organisation that secures the bulk of its work from major public and private sector clients may have to establish an EMS as a pre-qualification for tendering opportunities with that client. Some public and private sector clients, although currently a very small number, are beginning to look at environmental management in much the same way as they do towards quality assurance and health and safety. Clients are expecting the organisations that work for them to be more environmentally aware and, moreover, that any requirements they may set can be met.

Although an organisation may develop its EMS from QMS roots, it is appropriate that the two be kept separate with regard to third-party certification. This is essential in developing dedicated systems, documentation procedures and working instructions. Also, if difficulties should emerge in one system, they will not affect the certificated status of the other.

Environmental pre-qualification

At the present time, environmental management is not a general pre-qualification for any organisation to obtain work in the construction industry. However, it is well recognised that there is in the construction industry increasing concern for environmental protection and rising expectations of improved environmental performance by almost all organisations that contribute to the processes involved.

It is clear that environmental legislation will become more stringent. Organisations in all sectors of construction must become more aware of the need to identify, recognise and evaluate the effects of their business activities on the environment. The most appropriate way in which any organisation can meet the likely demands of the future is to develop and implement some kind of EMS, whether it be an internal or an external system. To be clear about the likely future evolutionary pattern of environmental management, one only has to look back at the development of total quality management (TQM) and the implementation of quality assurance (QA) in construction. In the early 1980s, quality assurance was little more that a new buzzword, but it has since become one of the most prominent, influential and accepted aspects of construction management.

If EIA is the environmental evaluation of a proposed construction project, then environmental management is the evaluation of the project's participants. In the years ahead, the link between EIA and the environmental contribution of the participants to the construction process may become established. This will mean that a line of environmental management might be expected throughout the total construction process. Should this occur, each participant will be evaluated and 'environmental pre-qualification' will become established.

Environmental management accreditation schemes

Environmental management accreditation schemes are already proliferating, mirroring the evolution of quality assurance schemes. The European Eco-Management and Audit Scheme (EMAS), formerly known as CEMAS, is a prominent example of this. EMAS is available to organisations involved in industrial applications, product manufacturing, services and product distribution and is site-specific. The basic premise of EMAS is to promote awareness of and empathy for environmental protection (environmental management) across the European Community. The scheme requires European member countries to develop and implement a structured eco-management and auditing process within their national regulatory frameworks. The UK equivalent of this is currently BS 7750 (ISO 14001), so in general terms the approaches are compatible. Companies which participate in the scheme, at present on a voluntary basis, are required to develop and implement an environmental protection system, which is in essence the equivalent of an EMS, in their organisations.

A similar scheme, SCEEMAS – Small Company Environmental and Energy Management Assistance Scheme – has existed in the UK since 1995. The aim of SCEEMAS is to make the EMAS scheme accessible to small and medium-sized enterprises (SMEs). The scheme is complementary to BS 7750/ISO 14001 and in general follows a system management approach to establishing and maintaining an EMS in the organisation.

Environmental management and the corporate organisation

Systems approach

To clearly demonstrate concern for and commitment to environmental issues, an organisation should establish and maintain an EMS. This system will ensure that the effects of an organisation's activities conform with environmental legislation and also with its own stated environmental policy and objectives and documented principles and procedures. The EMS should be a documented system of procedures and working instructions for practice and, moreover, the system must be seen to function effectively in implementation.

There is little doubt that any organisation will have some effect upon its environs, and the EMS must effectively and efficiently provide measures for

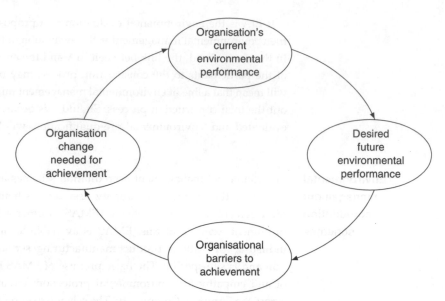

Fig. 6.2 Organisational development of an environmental policy

organisational interaction with its environment. An EMS must address a multitude of influencing variables, including not only the environmental aspect but also the socio-technical characteristics of the organisation. Environmental management systems, therefore, require a comprehensive framework of structure, procedures and practices and necessitate a wealth of organisational skills, abilities and commitment.

An EMS should be formulated such that it fundamentally seeks to prevent adverse effects from organisational activity rather than retrospective detection and cure. An EMS is a proactive organisational management system.

An organisation seeking to establish an EMS should ask three questions:

1. Why does the organisation need an EMS?
2. How does it develop an EMS?
3. What does the organisation seek to achieve through an EMS?

The object at this stage is to raise awareness of environmental issues in the organisation and begin to make proposals for policy and action. The organisation can develop an environmental management ethos through the formulation and implementation of a management system to incorporate all aspects of organisational activity. Such an approach aims to provide procedures and methods of working that either prevent problems arising or identify and manage problems in the most efficient and cost-effective way. The organisation will need to assess its current environmental performance, determine its desired future performance and relate these to the organisational change needed for success (see Figure 6.2).

A systems approach is especially applicable because:

- It is a 'whole organisation' philosophy and concept, and environmental management must be applied across activities to capture the mission and shared purpose of the organisation.
- It is a formally structured management approach to developing uniform organisational procedures to meet desired environmental performance.
- It presents a well-defined framework that structures the organisation's staff, skills and resources to meet the demands of environmental management in the most cost-effective way.
- It is a proactive rather than retrospective style that is essentially preventive, but also quickly reactive, in nature and can meet the dynamic characteristics required of environmental management.

The recognised advantages of using a systems approach are:

- identified framework and structure;
- uniform procedures and practices;
- documented evidence of compliance with performance criteria;
- improved communications and interdisciplinary efficiency;
- inherent preventive and reactive management capability;
- more rapid response to organisational problems or difficulty;
- effective accommodation of situational and organisational change;
- improved commercial competitiveness and marketability; and
- external recognition of organisational compliance with agreed system standards.

The priorities of an EMS should be to:

- increase awareness of potential organisational effects upon the environment within the organisation;
- take account of the specific environmental aspects interacting with the organisation and also broader environmental issues;
- identify potential environmental effects arising from organisational activity;
- meet current environmental legislation;
- enable organisation priorities to be determined with a view to creating organisational policy, objectives and goals;
- enable the planning, monitoring and control of organisational activity with regard to environmental effect;
- facilitate auditing, review and update;
- meet criteria of external regulatory/registration bodies; and
- allow sufficient flexibility for change to meet future circumstances.

Implementing such an approach will allow the organisation to produce a composite picture of its current business activities, and make appropriate policy and strategy to facilitate the transition to becoming environmentally aware.

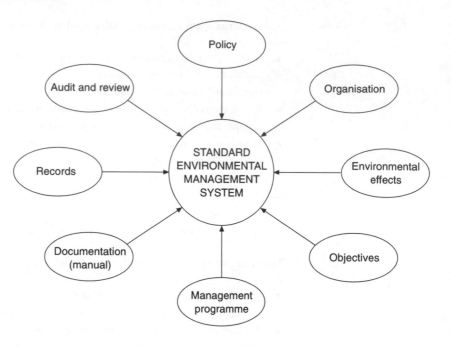

Fig. 6.3 Key elements of a standard environmental management system

Organisational elements of an EMS

To develop a systems approach to environmental management, an organisation will need to address the following elements (see Figure 6.3):

- preparatory environmental review
- environmental policy
- organisation and personnel
- environmental effects
- environmental objectives
- management programme
- documentation
- operational control
- records
- auditing
- review
- continuing environmental review and update.

Preparatory environmental review

The objective of the preparatory environmental review is:

> the detailed consideration of all aspects of an organisation's business with regard to its environmental situation as a basis for developing an environmental management system.
>
> (Griffith, 1994)

A preparatory environmental review is essential to determine the organisation's current environmental situation. The purpose of the preparatory environmental review should be to determine the current environmental status of the organisation with a view to developing its EMS.

A preparatory environmental review will involve:

- the detailed evaluation of existing environmental management procedures and working practices;
- an evaluation of environmental effects resulting from organisational activities;
- developing a register of the environmental effects of its business;
- determining the principal legislative and regulatory requirements affecting the organisation's business; and
- analysing organisational performance to identify areas for potential improvement.

Basic approaches to collecting information for the review can include the following:

- analysis of management structure and hierarchy;
- divisional and departmental analyses of activity;
- task analysis of activity;
- longitudinal and cross-sectional analysis of procedures and working instructions;
- method and work study of working practices and use of resources; and
- operational research of procedures.

A contemporary approach to such review is 'SWOT analysis'. This will allow the *s*trengths, *w*eaknesses, *o*pportunities and *t*hreats to the whole organisation to be considered.

Analysis and findings from the preparatory environmental review should be formalised in a report, which should focus on the following key issues:

- The determination of the nature and extent of problems identified and the organisational priority and time-frame for rectification measures.
- The development of an environmental management programme of action to address the issues raised in (1) (page 258) and how this will be resourced and managed.
- The required specification for the development of the organisation's EMS to address the aspects of (1) and (2) (page 258).

Environmental policy   The object of the environmental policy is to produce 'a published statement of organisational intentions in relation to potential environmental effects' (Griffith, 1994).

An environmental management policy should:

- define the organisation's corporate philosophy towards environmental management, in the context of its business activities; and
- be presented in the form of a policy statement, originating from the organisation's board of executive management.

The policy should:

- be relevant to the organisation's core activities;
- assess, genuinely, the environmental effects of the organisation's activities;
- be directed from executive level;
- be implemented at all managerial levels in the organisation;
- request commitment throughout the entire organisation and its workforce;
- make provision for setting, implementing and maintaining the environmental performance of the organisation; and
- be available to regulatory bodies and, more widely, for public scrutiny.

Although the structure and content of an environmental policy will vary according to the needs of the organisation, any organisation must ask itself a number of basic questions:

- What is the organisation's current environmental performance?
- What is the organisation's envisaged future environmental performance?
- What factors are assisting or preventing the organisation achieving its desired environmental performance?
- What aspects of the organisation must be changed such that it may eclipse its desired environmental performance?

In asking these basic questions, the organisation is not only setting out the basis for its environmental policy statement but also laying the foundations for its aims, objectives, strategies and procedures. Moreover, the need for organisational change will be addressed, and certainly the issue of environmental management policy making is one which is sure to challenge long-established organisational attitudes, customs and procedures. For this reason, if no other, environmental policy must come from executive level within the organisation and be seen clearly to come from that level.

The environmental policy should be published in an environmental policy statement, which is:

- clear, understandable and presented in a simple format;
- a true reflection of organisational intent and principles for action with regard to potential environmental effects;
- linked to the organisation's aim, objectives and goals;

- published with corporate identity; and
- flexible for use in organisational publicity material, such as the company's annual report and advertising literature.

Organisation and personnel

The object is to establish an organisational structure and set of management responsibilities to implement and support the EMS.

Organisation and personnel aspects of EMS development are concerned with three broad groups of activity:

1. Appointment of an environmental management representative.
2. Definition of responsibility, authority and interrelationships of environmental management personnel.
3. Organisation, communication and training of personnel involved with the EMS.

A management representative should be appointed by the organisation whose responsibility is for ensuring that the requirements of the environmental standard are met and the environmental management system is effectively implemented and maintained.

(BSI, 1992)

In an ideal situation, the management representative will assume exclusive responsibility for the EMS. It is more likely, however, that the management representative will assume responsibility for the EMS in addition to other organisational duties. In a larger organisation, an environmental management team may be constituted under the guidance of the assigned environmental management representative.

The role of the environmental management representative is to augment the traditional senior and middle management responsibilities and to co-ordinate environmental management activity throughout the organisation.

The environmental manager's remit extends to all aspects of the organisation's environmental management activity and therefore covers the organisation's relationships with external organisations, regulatory bodies and the public. The work of the environmental management representative relies upon the support of all line staff in all organisational disciplines, who in fact should maintain the environmental management system within the remit of their responsibilities. The challenge is to move the organisation from a position of low awareness to one of high awareness, acceptance and ownership.

In an EMS, the organisation should define and document in the environmental manual the responsibilities, authority and duties of those personnel who manage or functionally support the organisation's environmental management system.

To achieve the above, the organisation should:

- provide all necessary resources to sustain the system;
- clearly assign personnel to system duties;
- promote action to ensure compliance with the organisation's environmental policy;
- identify organisational problems and initiate action in response;
- ensure the capability to monitor, control, review and audit the system; and
- assign personnel to function in an emergency or disaster situation.

Within the organisation, management should determine the capabilities, skills, qualifications and experience of those persons who may be assigned to the EMS to ensure they can meet the explicit demands of the system. Organisation for the EMS should be arranged around such skilled and experienced personnel assigned to duties within the typical hierarchical framework of senior, middle and lower management. Where appropriate, job descriptions should reflect environmental management duties and performance criteria.

Good communication is essential to any organisation and to any organisational aspect, and it is also crucial to the development of the EMS. Communication through the organisational hierarchy should follow the usual good practices of short, succinct lines of communications allowing open and effective feedback routes, which are essential to environmental management review and auditing.

As an EMS may be a new approach for many organisations, training and education may need to be provided. This is needed for personnel at all levels of the organisation, as follows:

- for senior and executive management, to ensure that they understand the basis of an EMS so that they can generate policy, aims and objectives;
- for middle management, to ensure that they have sufficient training to develop and implement the EMS;
- for junior-level staff and personnel, to ensure that they can operate the EMS in practice;
- for new personnel, to ensure that despite turnover of staff the system can be maintained; and
- for existing personnel, to ensure that training updates accommodate new initiatives or changing circumstances.

Environmental effects   Environmental effects are the effects of organisational activity upon its environs, whether direct or indirect, detrimental or beneficial. The object is to implement an identification, assessment and documentation mechanism to address the organisation's environmental effects.

In an EMS, the organisation should develop, implement and maintain procedures for examining and evaluating the effects of its activities upon the environment. Classification of environmental effects by an organisation may include the following:

- emissions to the atmosphere;
- discharges to water;
- expulsion of solid, liquid and other wastes;
- use of land and natural resources;
- emission of other pollutants – noise, odour, dust;
- visual or comfort disturbances;
- contamination of land and other resources; and
- damage to balance of ecosystem structures.

Any of the aforementioned environmental effects may occur due to one of the following acts:

- the normal business and operating procedures of the organisation;
- operating procedures outside the usual or normal parameters of the organisation;
- accidents or catastrophes.

The level of detail of an environmental effects evaluation will depend upon a number of specific issues:

- the legislative requirements relating to the particular environmental effect and the relative situation of the organisation to the requirements;
- the nature and range of the organisation's activity and its environmental policy towards this activity; and
- the knowledge base within the organisation to determine and evaluate environmental effects.

In evaluating environmental effects, it is essential not only to determine the effect during normal operating conditions but also to anticipate the likely effect on the environment of abnormal activity accident or catastrophe. Risk assessment can be used to evaluate the potential degree of danger from the above situation, and depending upon the priorities determined, potential actions to mitigate effects can be considered. All information relating to the environmental effects of the business are documented in a register of environmental effects, a requirement of BS 7750/ISO 14001.

Environmental objectives

Environmental objectives are the measurable organisational targets or goals for environmental performance. The object for the organisation is to establish and maintain procedures to specify its objectives and measurable goals or targets at each management level in the organisation.

The objectives developed and goals set must be directed towards conforming with the stated environmental policy, and since they must measure performance achieved against an environmental plan, they should relate environmental performance to a set time-frame.

Because environmental management systems are proactive in nature and dynamic in their evolution, the objectives should demonstrate organisational commitment to continuous improvement and development. Objectives should be determined with due reference to the environmental effects evaluation and quantified where this is feasible for the development of performance goals or targets. Goals and targets set should be quantifiable and must be realistic, achievable and related to the activities of those managers and personnel responsible for particular aspects of the EMS.

Management programme

The environmental management programme is the organisational approach to formalising the means of achieving environmental policy, aims and objectives. The object is for the organisation to establish and maintain an environmental management programme. The programme must describe a number of essential characteristics, these being:

- the assignment of responsibility for environmental management functions within the organisation at various levels;
- a specification of goals or targets for performance at the different levels of management; and
- the way in which environmental management is to be achieved.

Any programme should identify and specify the following:

- the environmental objectives that are to be achieved;
- the means for achieving these objectives;
- the mechanisms for managing changing circumstances as encountered in implementation; and
- the corrective actions needed should difficulties arise in implementation.

The environmental management programme is fundamental to the establishment of the EMS as it is the core mechanism transferring organisational environmental policy into procedures and working practices. This is especially important with construction, where the ethos and policy of the corporate organisation must be successfully transferred to each individual construction project situation.

Documentation   The object for the organisation is to produce a set of documents that explain the EMS. The principal EMS document is the environmental management manual, which should describe the organisation's documentation of procedures and working instructions for implementing the environmental management programme. The manual serves as an organisation's reference point for implementation and maintenance of the EMS. In its simplest format, the manual can be likened to a organisation's general handbook of rules and working procedures but explicitly directed towards environmental management.

The nature of environmental management documentation can take many forms, depending on the organisation's business and circumstances, but generally follows a collection of written works:

- an EMS manual encompassing the whole organisation;
- supplementary manuals encompassing departmental or divisional activity within the organisation; and
- specialist manuals encompassing specific organisational tasks and functions.

While the precise nature, format and content of the environmental management manual will differ between organisations, the general sections forming the structure of the manual should be as follows:

- Contents list
- Revision list
- Distribution list
- Statement of authorisation
- Summary and instructions for use
- Policy description
- Organisational structure
- Applicable standards and regulations
- Aims and objectives
- Programme and procedures
- Goals and targets
- Operation and control mechanisms
- Record systems
- Review and auditing
- Training
- Application to 'third party' (subcontract) organisations.

Operational control   The organisation's objective is to specify mechanisms for control and actions with the EMS. Operational control within an EMS encompasses three organisational aspects:

1. Control mechanisms.
2. Verification, measurement and testing.
3. Non-compliance and corrective action.

The organisation should identify activities which have the propensity to affect the environment relevant to the organisation's policy and objectives. The organisation should plan and carry out such activities to ensure that they are conducted under controlled conditions. Control of organisational activity should consider:

- Procedures and working instructions contained within the organisation's environmental management manual.
- Criteria for environmental performance of the procedures and instructions.
- Monitoring and recording of activities within procedures.
- Procedures for managing 'third-party' (subcontract) organisations.

The EMS should develop and maintain verification procedures to ensure that organisational practices meet the specification of the environmental programme. Verification should be undertaken by system area, or specific activities, and, for each, the following tasks are pertinent:

- specify the verification information needed from the system;
- specify the verification procedures;
- specify performance levels expected to achieve verified practices;
- specify action to be taken if deficiency is identified during verification;
- specify that complete documentation is a prerequisite.

The nature, scope and detail of verification procedures should be contingent on the importance of the organisational aspect they review. Although it is not always possible, verification should be a measurable activity where actual performance can be assessed by measurement or testing. In such cases, the following should be undertaken:

- specify the measurement to be undertaken;
- specify the procedures to be used;
- specify acceptable performance;
- define the accuracy of the expected results;
- specify the recording mechanisms to be used.

The EMS should develop and maintain procedures for analysing activities which do not conform with the performance expected. For such instances, the environmental management representative or management team should:

- investigate the aspect suspected of non-conformance;
- determine the problem, its nature and extent;
- identify the cause of the difficulty;
- consider action to remedy the situation;
- initiate action;
- monitor and record attempts to rectify;
- adopt improved procedures;
- document revised procedures.

Structured investigation of suspected non-conformance in this way will allow the EMS to respond to deficient procedures and working practices in the most efficient and effective way and check system problems at an early stage to avoid prolonged detrimental environmental effects occurring.

Records The object is to maintain a set of records to demonstrate compliance with organisational policy and standards.

The organisation should develop and maintain a formalised set of environmental management system records. This is essential to demonstrate compliance with the system and also to facilitate evaluation of performance. EMS records should be maintained for all system aspects and should be stored such that they can be easily retrieved for use within the on-going system or to provide information for auditing and review purposes. The records system should be a composite part of the environmental management documentation, that is, it should be described in the environmental manual.

The organisation should develop a records system in which records are:

- maintained for all organisational system aspects;
- identifiable to activities;
- indexed;
- filed systematically;
- stored appropriately;
- maintained and updated;
- available for internal and external use.

Audit The object is to establish a procedure to facilitate auditing of the EMS. Environmental management audits are 'the periodic detailed evaluation of the organisation's environmental management system to determine its effectiveness in satisfying the environmental policy' (Griffith, 1994).

An EMS should develop and maintain appropriate procedures for conducting audits. Auditing may be internal for use in on-going review of the system or can be external for environmental assessment audits or certification purposes.

Auditing is essential to determine:

- that the EMS is following the organisation's management programme and is being implemented correctly; and
- that the EMS is fulfilling those requirements specified by the organisation's environmental policy.

To implement auditing effectively, the organisation should develop an audit plan, which is a systematic approach to determining the organisational aspects to be audited, frequency of auditing, procedures to be used, persons responsible and method of reporting.

Environmental audits may be carried out internally by the organisation's personnel, but to achieve impartial assessment external auditors may be employed. In addition, audits may be conducted by external control agencies, for example certification bodies and regulatory organisations.

Auditing should be the responsibility of the environmental management representative, who should initiate the audit, monitor its progress, collate information for analysis, distribute reports to all organisational levels and, when external auditors or other parties are involved, co-ordinate their activities.

Review   The object is to implement a procedure for system review. An essential aspect of the EMS is periodic management review. The organisation should ensure that management review of the system is conducted at appropriate intervals, first, to monitor the system and, second, to ensure that the system continues to meet the requirements of the standard and current legislation.

The nature and extent of the review should encompass the entire organisation and all its activities. Such review is the responsibility of the environmental management representative, who may assign internal personnel or external reviewers.

Management review should always follow environmental management system audits and should ensure that:

- Existing systems meet the current organisation's operational requirements.
- Any changes to legislation are incorporated in revised policy, objectives and procedures.
- All recommendations made in the audit have been implemented.
- The present system meets all prevailing circumstances in the wider context of organisational activity.

Environmental management review should form an integral part of system development and therefore should be documented like all other environmental management system aspects.

Continuing environmental review and update

The object is to develop a long-term review of the EMS. Because EMS is a proactive and dynamic management activity, the system is not simply complete once the eleven key organisational elements of environmental systems development have been satisfied. In fact, at this stage the environmental management system is merely up and running. The system has to be monitored, reviewed and upgraded as organisational experience is acquired. The environmental management system must therefore be regarded as an evolving organisational concept and be subject to continuing environmental review.

The importance of continuing environmental review is not merely the maintenance of the system but also the broader issues of environmental management across the organisation and in relation to the environmental situation that the organisation continues to find itself in. Change is an occupational hazard of almost all organisations. The organisation must not only keep itself abreast of the current situation internally but also continuously monitor and appraise outside influences, such as the changing environment, legislation, regulation, economic, political and market-oriented forces to ensure that its environmental management system continues to meet current demands.

For these reasons, continuing environmental review must be formalised as an 'overview' mechanism supplementing the management review of the system. The substance of the continuous review should also be formally discussed, analysed and actions considered at least annually and at other times where deemed necessary, to respond to changing circumstances.

Reviews of this type, addressing the broader issues of environmental management, may well precede the organisation's annual report; in fact, information generated in the continuing review may well form an important part of organisational reports to commerce, industry and the public. Continuing environmental review can therefore represent an important organisational mechanism in the public face of environmental management.

Environmental management and the construction project organisation

Environmental evaluation

The project evaluation stage

Environmental management during the project evaluation stage is manifest in the process of environmental impact assessment (EIA). This is the appraisal technique for ensuring that the potential environmental effects of any new development are identified and considered before any approval is given by the planning authority. EIA is a legislative requirement for specific construction projects in many countries. For example, across the European Union, the Environmental Assessment Directive (European Commission, 1985) underpins national EIA legislation in its member countries.

In the UK, the requirements for EIA are specified through the various national regulations in England and Wales, Scotland, and Northern Ireland. For construction projects requiring planning permission, the Environmental Assessment Directive is given legal effect in:

- England and Wales by the Town and Country Planning (Assessment of Environmental Effects) Regulations 1988 (HM Government, 1988a);
- Scotland by the Environmental Assessment (Scotland) Regulations 1988 (HM Government, 1988b);
- Northern Ireland by the Planning (Assessment of Environmental Effects) Regulations (Northern Ireland) 1989 (HM Government, 1989).

The principal function of EIA as an environmental management tool is to systematically provide as much information as possible on the potential environmental effects of a construction project, within the scope of applicable legislation. It must take into account inhabitants, fauna, flora, soil, water, air, climate, landscape, material assets and cultural heritage. In addition, it should appreciate the interactions between these elements.

The initiator of the EIA is the developer, who will play an important part in the process along with the following participants: specialist consultants; the planning authority; statutory and other consultees; the public; and, depending upon circumstance, central government departments.

The output from an EIA is the *environmental statement* (ES). This is:

a publicly available document setting out the developer's own assessment of the likely environmental effects of his proposed development, which he prepares and submits in conjunction with his planning application.

(DoE, 1989a)

An ES should provide information in five main sections (DoE, 1989b). These are:

1. Information describing the project.
2. Information describing the project site and its environs.
3. Information describing the assessment of environmental effects.
4. Information describing the measures to be taken to mitigate the effects.
5. Information describing the risks of accidents and hazardous development.

Determining the need for environmental impact assessment

There are five situations where the planning approval process will consider the relevance of environmental assessment. These are:

1. A developer applies for planning permission with no reference to EIA because it is not necessary or the developer does not think it is necessary. In the latter case, the planning authority will advise on the need.

2. A developer applies for planning permission and submits an environmental statement because it is necessary under the EC Directive and therefore lies within national planning regulations.

3. A developer applies for planning permission and asks the planning authority if an EIA is necessary because he is unsure if the proposal falls within the scope of regulations. In this case, the planning authority will advise on the need.

4. The Secretary of State may independently exercise power to request an EIA where, for example, representations are made by third parties about a proposed development.

5. The planning authority must undertake an EIA before granting itself permission for development, because the proposal falls within national planning regulations criteria.

Outline elements of EIA

There are eight elements in the process of EIA (DoE, 1995). These are:

1. *Project description*: a clear description of the project and its works.
2. *Screening*: the process of determining the need for environmental assessment.
3. *Scoping*: directing the assessment towards aspects of particular importance.
4. *Baseline studies*: the identification of significant environmental effects.
5. *Impact prediction*: the consideration of the degree of impact.
6. *Mitigation assessment*: the measures to be taken to mitigate the effects.
7. *Environmental statement*: a publicly available environmental effects statement.
8. *Environmental monitoring*: monitoring of environmental effects on the project.

It is not intended within the scope of this chapter to describe the process of EIA. Other works cover this important process (Glasson *et al.*, 1994; Griffith, 1994). However, it is important to stress that considerable information is acquired from this EIA process which feeds into the briefing, design and contract administration stages of the construction project.

While there is considerable documented evidence both in support and critical of EIA regulation and procedures (Coles, 1992; Coles and Fuller, 1990; CPRE, 1992; Jones and Wood, 1991; Weatherall, 1992), the principal limitation to successful implementation is the severely limited scope of legislative jurisdiction. Although under review, current criteria determining the need for EIA in construction planning and application processes are restricted to major industrial, infrastructural and development projects. The vast majority of construction projects do not come within the criteria specified, and therefore EIA is not currently mandatory. In these circumstances, EIA is left to the goodwill and intent of the developer or client, and given this one cannot blame these parties for not subscribing voluntarily to environmental safeguards, particularly in the tight economic climate prevalent at this time.

Contribution of EMS to project evaluation

While many construction projects do not fall within the legislative criteria for EIA, there is a growing trend among developers and clients to undertake EIA on a voluntary basis. Many organisations see this as a positive projection of their commitment to and actions towards environmental protection. Such developers commission EIA to inform the briefing, design and construction processes and to ensure that the resulting development represents the best environmental option that is practically achievable. Organisations may find that such an approach fits in well with their current commitment to the environment, which is likely to be displayed through its EMS.

An organisation that establishes an EMS will be in an ideal position to undertake EIA of its development projects, as it will, through application and experience, have already identified the environmental effects of its business and needs only to focus upon the environmental effects of the project site. Organisations which are, and are seen to be, environmentally empathetic will find that they can respond much better to the requirements of the planning process than those which are not. They are also seen by the regulatory processes and the outside world as organisations which have regard and respect for the environment. EMS gives the organisation the framework, structure and rigour to understand EIA effectively, which helps the process itself but moreover allows the organisation to see the project better in relation to its wider activities.

Many organisations also find that EMS is essential to environmental monitoring. Environmental monitoring can benefit a developer in a number of ways. First, any deviation from the predicted EIA can be seen at an early stage, and therefore management can respond quickly and more effectively. In this way, monitoring is developed into an auditing and review mechanism, which is the likely long-term development of the EIA process. Second, environmental monitoring is a useful source of information for the developer. It may be used to defend claims for environmental damage, for verification purposes in regulatory inspections, and to support the organisation's activities in certification schemes. Third, environmental monitoring allows the organisation to see the longer-term environmental picture, and information generated from one EIA and monitoring initiative can be used purposefully on the next project, and so on, leading to an environmental database with the organisation.

Those larger organisations which have an EMS in place are undoubtedly likely to use it as a basis for procuring their projects. It can give them the necessary management rigour better to control the development, briefing, design and construction processes and can be easily extended to form the basis of operating procedures and working practices in the use of the finished development.

An EMS can benefit the developer when conducting an environmental assessment in the following ways:

- By producing the necessary outline, framework and structure in the organisation to get the best from EIA.
- In-house expertise, where available, may assist the appointed team of environmental consultants to conduct the EIA.
- The environmental effects of the developer's past and current environmental performance will be known.
- There is a lesser focus on the developer and more attention given to the project site and its environs.
- Information is gathered over the longer term through environmental monitoring, which provides a wider range of knowledge than might otherwise be available and information which should be more reliable.

This will lead to better EIA and EMS as experience is gained.

The briefing and design stages

The briefing stage

Briefing is a core activity that begins the construction process in real terms. An accurate and effective brief is essential for, first, translating those environmentally sympathetic aspects identified in EIA and, second, for laying the groundwork for an effective design phase. Briefing, in a conventional construction situation, can be a difficult task, and it is made all the more problematic when environmental demands are additionally placed upon it. The principal cause of difficulty in the briefing process is invariably one of information provision and communications management – more specifically, the difficulties of handling large amounts of information coming from many different participants simultaneously; the correct interpretation of communication between the parties, even the evolving pattern of the briefing process, does not lend itself to accurate perception and projection of project information. These aspects in themselves are difficult to manage and are made all the more fraught as environmental management aspects, which are at this stage only anticipated and not confirmed, are postulated for consideration in the briefing process. The client is essentially 'at the helm' in the early part of the briefing phase. Unfortunately, the client is not involved solely with this process but is rather more extensively involved in a great many other project interests, which may deflect him from the important task of briefing. An effective final brief will be dependent on the many participants in the process integrating the knowledge gained from the EIA and their contribution to project development (see Figure 6.4).

EIA may have highlighted particular environmental aspects which could be significant to project briefing and design. Such aspects must be successfully incorporated into the information which is generated during these two important stages.

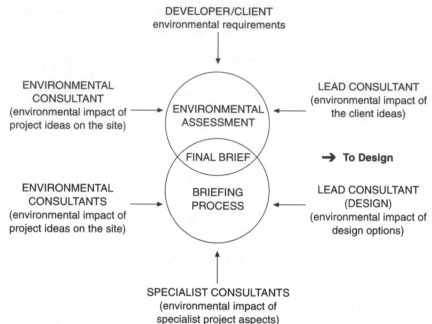

Fig. 6.4 Contribution of knowledge to the final project brief

The principal aims of the briefing stage, with regard to environmental management, are:

1. to accurately translate the potential environmental effects of the project which were identified in the EIA into measures of mitigation within the brief and design;
2. to ensure that the environmental aspects identified are balanced with the expectations of the client and the project criteria; and
3. to provide the design team with a clear set of environmental wants, for example situation, form, function, layout, style, performance.

Experience (Griffith, 1994) has shown that it is vital to eliminate key problems concerned with environmental consideration during the briefing process. Key problems identified are:

*   Environmental concepts may not have been adequately identified and considered during EIA, leading to a lack of information for use in briefing.
*   The client may not fully appreciate the potential for environmental effects and therefore not appreciate their impact on briefing and design.
*   The client may lose sight of environmental orientation in relation to other project criteria such as time, cost and performance.
*   The client may assume a passive role in environmental safeguards where proactiveness by the employer would be desirable.

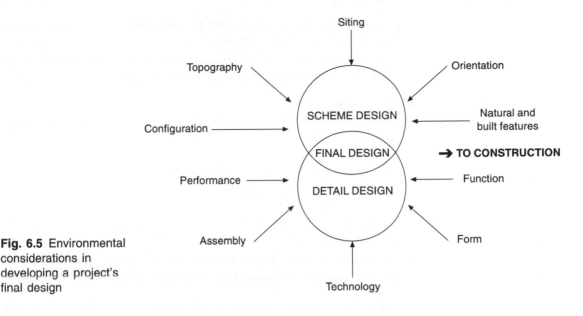

Fig. 6.5 Environmental considerations in developing a project's final design

The client can play a significant part in encouraging environmental management in the briefing stage by:

- *during feasibility studies*: identifying clearly the environmental orientation of the project in relation to the aims, objectives, resources and management.
- *in the outline brief*: emphasising the environmental orientation established during EIA.
- *in the appointment of consultants*: appointing specialist (environmental) consultants to assist the principal design/engineering consultants.
- *when ordering procurement method*: choosing a method that best meets the environmental orientation.
- *in cost assessment*: budgeting for environmental management.
- *during site investigation*: consolidating the findings and suggestions of EIA.
- *in the final brief*: drawing together the environmental information and balance with other project criteria, such as time, quality and safety.

The design stages
The design stage, traditionally divided into two key elements, (1) scheme design and (2) detailed design, lends itself to environmental consideration (see Figure 6.5).

1. *Scheme design*: is concerned with the consideration of the evolving construction form, addressing such aspects as position, layout, shape, size and material constituents, i.e. those aspects that make the product aesthetically and situationally acceptable.

2. *Detailed design*: is concerned with the consideration of technology and structure to create the built form, addressing such aspects as components, materials and the assembly processes, i.e. those aspects that allow the product to perform and make it technologically acceptable.

Scheme design
The principal tasks of scheme design, with regard to environmental management are:

1. to translate the brief into a technical specification for materials, products, components and construction processes, taking into account environmental requirements.
2. to ensure that environmental needs and expectations are met while satisfying the overall requirements of the project.
3. to assign responsibility to consultants for the various specialist inputs, including aspects for environmental consideration.

Although all construction projects are different and must therefore be considered on their own merits, scheme design should consider some or all of the following environmental aspects, many of which will have been investigated as part of the EIA.

1. *Effects on land*
   - landscape
   - topography
   - stability
   - soil constitution
   - mineral resources
   - soil disposal
2. *Effects on water resources*
   - natural drainage
   - groundwater level
   - watercourses
   - underground water
   - pollution aspects
3. *Effects on atmosphere*
   - emissions
   - odours
   - noise
   - vibration
   - reflections
4. *Effects on ecology*
   - flora and fauna
   - natural habitats
   - ecological balance

5. *Effects on population*
   - density
   - proximity
   - intrusion and disturbance
   - population change
   - infrastructure
   - amenities.

Scheme design is also concerned with the layout and configuration of the internal aspects of the project, as in the case of a building for example. Scheme design, therefore, should address internal allocations of space, space utilisation, occupation and usage.

Scheme design will call upon that information generated by the EIA during the evaluation phase, supplemented by an environmental site survey conducted by the environmental consultant and designer. This will ensure that the potential environmental effects of the project are identified and considered before the client is committed to the detail of the project. Scheme design study provides useful information to both the scheme design and detail design and is an important part of the on-going design review procedure as the information gathered at this stage reinforces aspects considered in the brief.

Periodic reappraisal of the scheme design is essential in order to ensure that the design continues to meet all the specified requirements and client needs. There are always likely to be changing aspects in design as more information is generated and applied to the project situation. Contract review is therefore essential to the early design processes. It is not uncommon for the brief to be weak in some cases, and this can go undetected until information is reviewed during scheme design. However, it is likely that only such weaknesses in the scheme design resulting from an inadequate brief will be the result of deficiencies in the original EIA.

Likely problems to be encountered in the pursuit of environmental management during scheme design are (Griffith, 1994):

- EIA can be very subjective, and the potential environmental impacts on the project can be difficult to anticipate fully.
- Unless firm environmental orientation is established, it is easy for aspects to be changed as scheme design translates to detail design.

Within scheme design, the principal consultant can make a significant contribution to environmental management and alleviate potential problems by:

- in reviewing the brief ensure that all environmental information is clear and available to the scheme design consideration;
- in the scheme design study consolidate EIA with a detailed environmental site survey (ESS);

- in the scheme design review ensure that any matters outstanding from briefing are highlighted and considered; and
- in finalising scheme design ensure that all environmental aspects are translated into considerations to be given in detailed design.

Detailed design

The principal tasks of detailed design, with regard to environmental management, are:

1. to translate the scheme design specifications into detailed and workable design concepts, including any environmental aspects that have been identified.
2. to ensure that the above are accurately represented in the contract documentation.

There are four main phase tasks within detailed design formulation. These are:

1. Design study.
2. Consideration of alternatives.
3. Drawing up and documenting.
4. Verification and final design.

In considering environmental issues within these tasks, the designer should undertake the following:

- Develop those environmental ideas generated in the scheme design towards fully detailed design elements.
- Investigate a 'green' or eco-design and construction approach as an integral aspect of the design task.
- Ensure that detailed design leads to comprehensive and complete project documentation, specifications and drawings incorporating specified environmental performance criteria.
- Consider all statutory legislative requirements relating to environmental management in construction (health and safety, pollution, land use, etc.).
- Evaluate eco-materials and components for potential inclusion in the final design.
- Consider value for money, balancing environmental management with other project priorities.
- Establish detail design review procedures to appraise design progress on an on-going basis.
- Control the various interfaces within the design process, i.e. client/ designer, designer/environmental consultant, etc.

The detailed design phase integrates and consolidates the scheme design through the production of detailed design drawings, specifications and contract documentation. During the detailed design phase, the philosophy towards environmental management is clearly established through developing an environmental management design plan that takes into account the basic environmental position of the client and extends this to address how environmental management can become an integral and practical part of the construction phase.

Likely problems encountered in considering environmental management during detailed design are:

- Little information may be available concerning 'eco-specification', i.e. selecting those materials and components which are environmentally friendly.
- New and recently amended legislation in environmental aspects is not widely understood and interpreted correctly.
- The user, who realistically determines the fitness for purpose of the end-product, is often not consulted during detailed design.
- Features incorporated into the detailed design are not thought through to the construction stage, i.e. little consideration is given to environment-related buildability/constructability.

In detailed design, the principal consultant can make a significant contribution to environmental management by:

- *during detailed design study*: ensure that client and user environmental requirements are met.
- *in considering design alternatives*: determine environmental options and balance equally with other project requirements.
- *in detailing drawings, specification and contract documentation*: clearly interpret and present environmental needs and eco-specify where appropriate.
- *in finalising detailed design*: ensure that the design meets all environmental needs before committing to construction stage.

Environmental control systems

The contract administration phase

Environmental management during the contract administration stage is chiefly the responsibility of the main (principal) contractor, overseen by the client's lead consultant or other appointed representative based on site.

If the contractor has an EMS within its organisation then this will have already established the situation for environmental management in the construction process, and therefore the primary aim for the contractor will be to transfer the corporate system concepts to management mechanisms on site by which environmental measures can be established. The site structure and organisation of the contractor's EMS should present the formal procedure within which the environmental protection measures demanded by the client can be achieved.

Pre-construction

During the pre-construction phase, between tender and commencement on site, the contractor should undertake the following:

- Review carefully the client's environmental management requirements as specified in the contract documents.
- Attend pre-contract meetings and address specifically all aspects of environmental management relating to the site processes.
- Appraise the contractor's own corporate EMS in relation to those aspects identified and begin to translate the corporate EMS into a project-based environmental programme.

Construction

A number of these aspects relate to little more than good, sound site practice, but they are nonetheless essential to effective environmental management at project level. The contractor should ensure that he:

- provides a well-structured and organised approach to site organisation;
- clearly determines the requirements of environmental management and assigns responsibility to site personnel;
- constantly monitors environmental management aspects on site in a systematic way and records all environmental issues;
- constantly liaises with the client/designer/environmental consultant to ensure continued sound environmental practices; and
- provides a formalised system of deliveries for environmentally sensitive resources, including appropriate storage, protection and distribution on site.

Contractor's EMS and project environmental management programme

The essential aim for the contractor is to embody the EMS of the corporate organisation into a site-based project EMS. This involves the establishment of an environmental management programme that focuses on identifying environmental risks associated with the works, assigning responsibilities for mitigation risk, and implementing mechanisms to record and review any occurrence of risk when it occurs. It is important that the contractor develops simple mechanisms for site-based environmental management. A problem can be that too complex an approach is established where site management becomes consumed in a laborious and bureaucratic paper-based systems. The basic EMS framework is set by the corporate organisation, and therefore the project

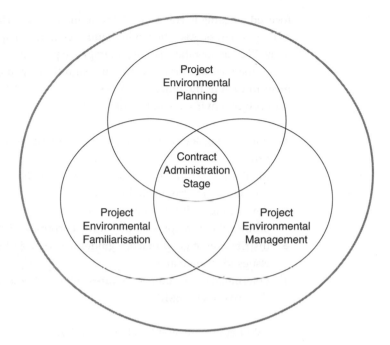

Fig. 6.6 Functional areas of contract administration in developing a contractor's environmental management programme

is concerned with implementing that system through site responsibilities and mechanisms that are as simple as possible.

Establishing a project-based EMS through an environmental management programme involves the main contractor in three functional areas of management (see Figure 6.6).

1. *Project environmental planning (EMP)* – concerned with environmental consideration during the tendering process and the pre-contract process up to commencement on site. Prominent tasks are risk identification and assessment and environmental planning based on the corporate EMS.
2. *Project environmental familiarisation* – concerned with environmental consideration at commencement on site. Prominent tasks are briefing staff, site orientation and awareness training.
3. *Project environmental management* – concerned with environmental consideration during the administration of the production process on site. Prominent tasks are communication of procedures and implementing procedures.

These stages should be considered such that they provide systematic identification, evaluation and management of key potential environmental effects, for example atmospheric emissions, discharges, spillages, noise, vibration, dust, fumes, waste and hazardous substances. Activities within the EMP should provide feedback which is compatible with the EMS of the corporate organisation. This is important, since corporate management will be multi-project focused, needing to collect feedback on many and varied construction projects as they progress. Also, they will require feedback to appraise the effectiveness of the EMS in operation, seeking to improve procedures as experience is gained.

The main contractor can contribute significantly to environmental management in each of the functional areas:

In project environmental planning by:

- undertaking an environmental risk assessment as part of the tendering process;
- identifying key environmental effects that will need to be managed on site through an environmental site survey (ESS);
- developing a documented environmental management plan (EMP) that follows the principles set out in the corporate EMS;
- distribution of good practice guidelines to project staff relevant to the key issues identified; and
- determination of audit procedures between the corporate organisation and the project site (EMP).

In project environmental familiarisation by:

- briefing all staff on the environmental impact issues identified in project environmental planning;
- conducting a site tour to familiarise the project team with the site and its relationship to the areas of environmental effects; and
- providing training in the use of procedures that will be used to plan, monitor and control environmental effects.

In project environmental management by:

- providing practice notes to site staff on environmental management procedures to be adopted;
- using an item checklist for environmental risk assessment and risk monitoring;
- developing self-audit/review sheets for use by EMS managers;
- providing guidance notes on potential actions should environmental problems arise; and
- specifying designated senior (corporate-level) managers who can be consulted easily and quickly should problematic environmental matters arise that cannot be resolved at project site level.

The contractor's ESS should systematically identify to management those aspects of site activity which will have a significant potential to affect the site and its environs. These include, for example, spillages, discharges, contamination and forms of pollution. Once identified, the contractor will be in a position to plan, monitor and control for those environmental effects to ensure that they are prevented, or where they do occur, are minimised, i.e. implement the EMP on site.

Considerable information will have come from the contractor's pre-tender site visit and, in addition, information should be available from the client's EIA report and consultant's design data. All these sources should be reviewed before commencement on site. Some of the potential environmental effects will be subject to legislative control, and therefore the contractor is bound to mitigate their manifestation. Further aspects will be covered by the client's specification requirements and therefore must also be actively addressed.

Other aspects should form part of the contractor's voluntary attempt to protect the environment in the course of good site practice. Where this latter aspect is concerned, much is dependent upon the discretion of the contractor and the quality of his construction practices.

The contractor's environmental management plan

The contractor's ESS will have identified the potential for environmental effects on site from construction activity and brought them to the attention of supervisory management and the workforce. They should then transfer and apply these principles in the site-based processes through the project EMP. This plan should develop its approach through three principal areas of consideration:

1. Specify requirements
2. Assign responsibilities
3. Implement mechanisms.

Requirements

In each area of site activity the contractor should ensure that:

- The potential for environmental effect is clearly identified (risk identification).
- The magnitude of the environmental effect is determined (risk assessment).
- There are planned controls to prevent the environmental effects occurring (risk mitigation).
- There is a course of action to 'manage' the environmental effect should it occur (risk management).
- There is an active monitoring procedure to indicate if harmful environmental effects occur.
- All environmental effects are systematically recorded, together with any measures taken in mitigation.
- Environmental situations are reviewed to determine cause, the effectiveness of actions taken and the project environmental management plan updated.

Responsibilities

It is the responsibility of supervisory management to ensure that the above requirements are met by employing systematic mechanisms that:

- make all personnel aware of the potential for environmental effect and impress upon them the need to report any breach of environmental integrity;
- ensure recorded daily checks in all vital areas of site activity;
- review daily checking procedures at weekly site meetings;
- collate weekly data for discussions at monthly project meetings; and
- ensure that information is collated for long-term performance review.

Mechanisms

To assist the practical functioning of the EMS on site, simple recording and reporting mechanisms should be used. Pro forma checklists can assist to formally record the following:

- occurrence of any environmental effect
- location of the incident
- reason for the occurrence
- any action taken
- review of action to assess effectiveness
- further action needed
- notifiable nature of specific incidents, for example a report to authorities of breaches of environmental regulations
- feedback to project files and corporate records for auditing purposes.

Environmental management system development checklists and outline framework

This section presents a series of checklists and an outline framework to assist in the basic conceptual development of an environmental management system. For further information on environmental evaluation and control systems for application to construction industry organisations, the reader is directed to the references section of this chapter.

It is not intended to provide detailed pro formas for system implementation in this chapter. However, a series of suitable pro formas can be developed for environmental aspects following the comprehensive details and pro formas presented in Chapter 5.

Table 6.2 Checklist of factors to consider in determining the environmental position of the organisation

External Factors:
- Regulation and legislation
- Standards
- Insurances
- Marketplace
- Customers
- Competition
- Finance sources

Internal Factors:
- Environmental performance
- Awareness and knowledge
- Structure and organisation
- Management skills
- Staff
- Resources
- Risk

Table 6.3 Checklist of key steps in developing the environmental approach

- Obtain executive commitment
- Determine framework, management and organisation
- Undertake preparatory environmental review
- Establish policy
- Create action plans
- Document procedures
- Communicate policy, plans and procedures
- Implement plans and procedures
- Determine success of implementation
- Review and improve total process

Table 6.4 Checklist for considering the preparatory environmental review

Phases:
- Thinking
  determine scope and objectives of the evaluations
- Planning
  determine plans, organisation, methods and resources
- Implementing
  gather information, collate, evaluate and report
- Reviewing
  determine actions to be taken as a result of the total review

```
Version Control:
First published: 24 March 1995
This revision: No. 5, 27 June 1997
```
Statement by Executive Board:

ABC International Management Ltd provides a wide range of services to the construction and related industries. In so doing, it acknowledges the need to safeguard the environment in all aspects of its business. The overall aim of our business is to provide complete customer satisfaction and where possible ensure that the services we provide present minimal detrimental effect to the environment and both the customer and ourselves.

In all our activities, we commit:

- to comply with all legal requirements and standards associated with environmental management.
- to review through our environmental management system all our activities and actively seek improvement.
- to develop and market services that have excellent environmental characteristics.
- to promote environmental excellence among our subcontractors, suppliers and other employed organisations.
- to, where possible, implement resources to safeguard the environment and its population.
- to be pro-active to environmental concerns.
- to ensure that corporate policies are implemented through sound environmental procedures in all our divisions and property organisations.

Fig. 6.7 Guidelines for the development of an environmental policy statement

Table 6.5 Checklist for prioritising organisational activities and their aspirations

	Activity	Target aspiration
1	Develop environmental policy for the organisation	Fundamental goal
2	Develop environmental policy statement for internal and external circulation	Fundamental goal
3	Identify organisational areas for attention and focus for management systems	Fundamental goal
4	Publish environmental management system manual, guidance notes and procedures for areas identified in 3 above, translate procedures into management assignments, and action plans	Improvement
5	Determine knowledge and skill base needs and implement training programme and review in relation to system procedures used	Improvement
6	Evaluate system processes to provide better procedures, expand meaningful experience and feedback and review	Greater improvement
7	Evaluate training programmes (in 5) and action plans (in 4) to determine improved performance	Greater improvement
8	Produce report on systems operation and documentation	Greater improvement
9	Develop and implement audit mechanisms	Greater improvement
10	Review and restate policy, guidance notes and documentation for whole system to minimise detrimental environmental impact of organisational activities	Ultimate goal

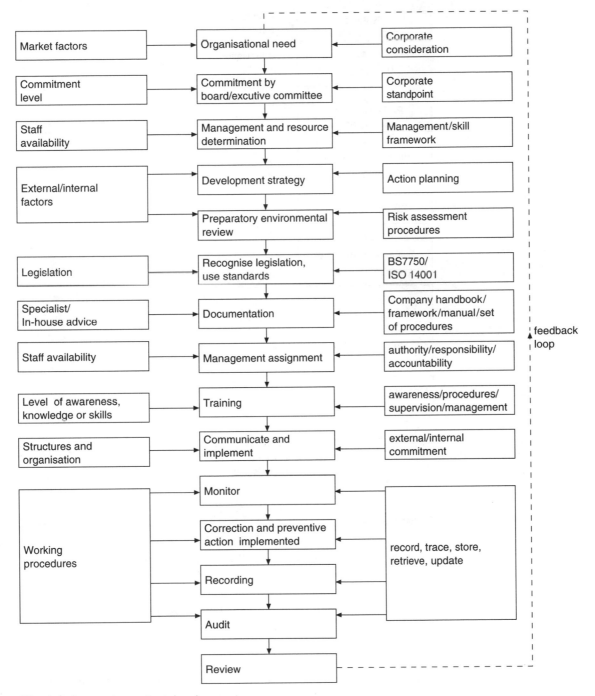

Fig. 6.8 Corporate system development

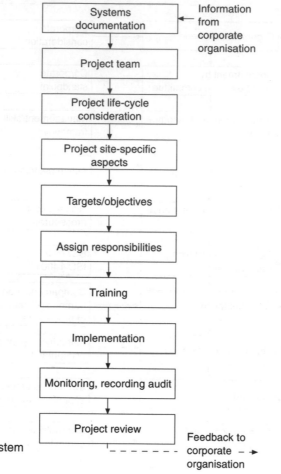

Fig. 6.9 Project system development

References

British Standards Institution (BSI) (1992) *BS 7750: Specification for Environmental Management Systems*, BSI.

Coles, T. (1992) *Practical Experience of Environmental Assessment in the UK*. Technical Report, Institute of Environmental Assessment (IEA).

Coles, T. and Fuller, K. (1990) *Analysis of the Environmental Impact Assessment Market in the UK*. Technical Report, Institute of Environmental Assessment (IEA).

Construction Industry Research and Information Association (CIRIA) (1995) *Effective Environmental Management*: Discussion Notes, CIRIA.

Council for the Protection of Rural England (CPRE) (1992) *Mock Environmental Directive*, CPRE.

Department of the Environment (DoE) (1989a) *Environmental Assessment: Explanatory Leaflet*, HMSO.

Department of the Environment (DoE) (1989b) *Environmental Assessment: A Guide to the Procedures*, HMSO.

Department of the Environment (DoE) (1995) *The Eco-Management and Audit Scheme*, DoE.

European Commission (1985) *The Environmental Assessment Directive on the assessment of the effects of certain public and private projects on the environment (85/337/EEC)*, HMSO.

Griffith, A. (1994) *Environmental Management in Construction*, Macmillan, Basingstoke.

Griffith, A. (1995) *EMS – Environmental Management Systems: An Outline Guide for Construction Industry Organisations*, Hong Kong Polytechnic University.

Griffith, A. (1997) *Environmental Management in the Construction Process*. Construction Papers No. 75. The Chartered Institute of Building (CIOB).

Glasson, J., Therivel, R. and Chadwick, A. (1994) *Introduction To Environmental Impact Assessment*, UCL Press.

HM Government (1988a) *The Town and Country Planning (Assessment of Environmental Effects) Regulations*, HMSO.

HM Government (1988b) *The Environmental Assessment (Scotland) Regulations*, HMSO.

HM Government (1989) *The Planning (Assessment of Environmental Effects) Regulations (Northern Ireland)*, HMSO.

International Standards Organisation (ISO) (1994) *ISO 14001: Specification for Environmental Management Systems*, ISO.

Jones, C. and Wood, C. (1991) *Maintaining Environmental Assessment and Planning*. Technical Report, Institute of Environmental Assessment (IEA).

Weatherall, V. (1992) *Operation of the UK's Environmental Assessment System: Is There a Need for Modification?* BSc degree thesis (unpublished), Heriot-Watt University, Scotland.

7 Information technology and communications systems

Introduction

Information systems and communications represent an important part of the construction process. This chapter looks at information technology generally and how systems, in particular, can be used for the generation and communication of information. Specific types of information system and software are identified in relation to the stages of construction. Information technology is also viewed from the perspective of strategic relevance to construction organisations. Several issues are considered, together with costs and benefits of using information technology in a construction organisation.

More in-depth consideration is given to the analysis and development of information systems and the different approaches available. The generation of information from developed systems is further addressed, together with the coverage of communications and networks. Telecommunications systems are identified, in addition to different types of network that can be installed and used by construction organisations. This is further extended by looking at accessing information on a global scale through the internet and considers remote access, including issues of network privacy and security.

From development and communications, the implementation of information systems is addressed, together with the consideration of hardware and software, and implementation methods for information systems. An important part of implementation is the consideration of human–computer issues, and in particular, the education and training of staff who operate and use information systems in the workplace.

Management issues that need to be addressed for effective and efficient operation are also outlined. Finally, consideration is given to possible developments in information systems that will have an influence on construction organisations and construction processes in the future.

Information in the construction process

Information and its effective and efficient communication are an integral part of the construction process. The consideration of a construction product

involves several stages from design and construction to operation, maintenance and demolition. The parties involved during these stages, particularly
with regard to design and construction, would typically include clients, professional advisers and consultants in addition to contractors, specialist contractors and suppliers. The types of information and extent of detail would
clearly be related to the project and the parties involved but would typically
relate to administrative, technical, financial or legal information. The increasing complexity of buildings, methods of procurement and the involvement of
parties providing various services, materials and components necessitates the
effective and timely communication of information.

From a contractor's point of view, information can be considered as both
internal and external to a company, as shown in Figure 7.1. Internal information relates to information which is communicated within the company environment to other departments, offices and project sites where construction
work is taking place. External information relates directly to other parties
during the various stages of project development, the extent and type being
dependent upon the project and type of contract.

Information management and the information systems in place in organisations are established and operated with varying degrees of sophistication.
Much is dependent upon the size of an organisation, its trading activities,
specialisation of work and other parties involved. Many companies have
inherited or developed systems which have been in place for years. Many
systems in the construction industry are still paper-based. Some organisations
have computerised some part of their systems for information generation and
communication, and others are striving for paperless systems and have a
vision of total computerisation for data capture, storage, use and communication. Certainly, the information requirements in construction organisations are
considerable owing to the increasing complexity of buildings, and legal, statutory and contractual requirements.

The economic climate and the cyclical nature of the industry often mean
that construction organisations are continually changing through expansion,
horizontal and vertical integration, and often, coerced diversification of trading activities. Thus, the information requirements of organisations may also
undergo change, and information systems should therefore be designed to
cope with the increased influx and diversification of information requirements.
Effective information management therefore becomes paramount.

Information technology in the construction industry

Advances in information technology (IT) in construction are now proceeding
at an alarming pace. Certainly, developments in technology are no longer the
limiting factor in the industry's reluctance to embrace technologies which

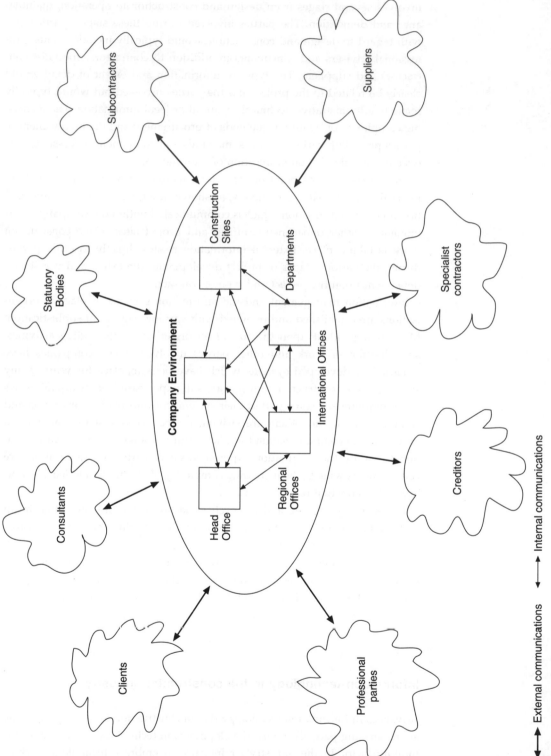

External communications ↔ Internal communications ↔

Fig. 7.1 Communications in relation to a company

other industries have been using for some time. Previous arguments about the fragmentation of the industry, different professional parties with different technical and cultural backgrounds, while still problematic, should no longer be used as an excuse. Powerful computer processing facilities are now affordable by all, and telecommunications systems are becoming more sophisticated and reliable. Indeed, commitment and the desire to use today's technology by individuals and companies are the major influencing factors.

Applications software is now used for a wide range of activities in addition to generic software for administrative functions (CICA, 1998; CIOB). The annual publication of the CICA software directory indicates a comprehensive classification of applications software, and no doubt this will continue to grow as technology advances. Software is now available which covers the whole construction process and can, to a large extent, be directly related to the stages involved, as shown Figure 7.2. Moreover, software houses are conscious of the need to share data between functions, and some products provide links to other software.

The focus on integration of data for construction processes has been around for some time, and research will no doubt continue in this area to provide total integration of information throughout the industry.

Technological advances have also led to the emergence of sophisticated multimedia technology. This technology provides a rich source of information representation through different and integrated media modes. Additionally, telecommunications and network management have provided for remote and real-time communications, in addition to group support systems, groupware and video-conferencing. These developments have provided a more efficient and supportive working environment. However, such developments have not been restricted to the offices of professional parties. Technological advances have also been made in connection with the production process, which has greatly influenced construction operations.

Innovative technologies have already been introduced on construction sites and have been applied in a variety of ways (Nunn, 1997). This includes electronic handling of drawings, 3-D modelling and virtual reality applications. Data capture techniques have included wearable equipment for video representation and site surveys, in addition to voice-activated data capture incorporating radio modems beamed back to site offices to deliver real-time data. Systems are also used for proposed changes to a project, and variations can be analysed quickly to determine the cost-effectiveness of changes, allowing decisions to be made at each stage of a project. This form of technology will no doubt have an impact on sites in the future, providing more co-operative and flexible working (Nunn, 1997). While the technology available can address many of the previous problems experienced, the extent of computer use across the industry is still somewhat varied. In the main, the use of IT and computers is determined by companies themselves. It could be argued that the nature of the industry itself still has an influence. The industry has the distinction of

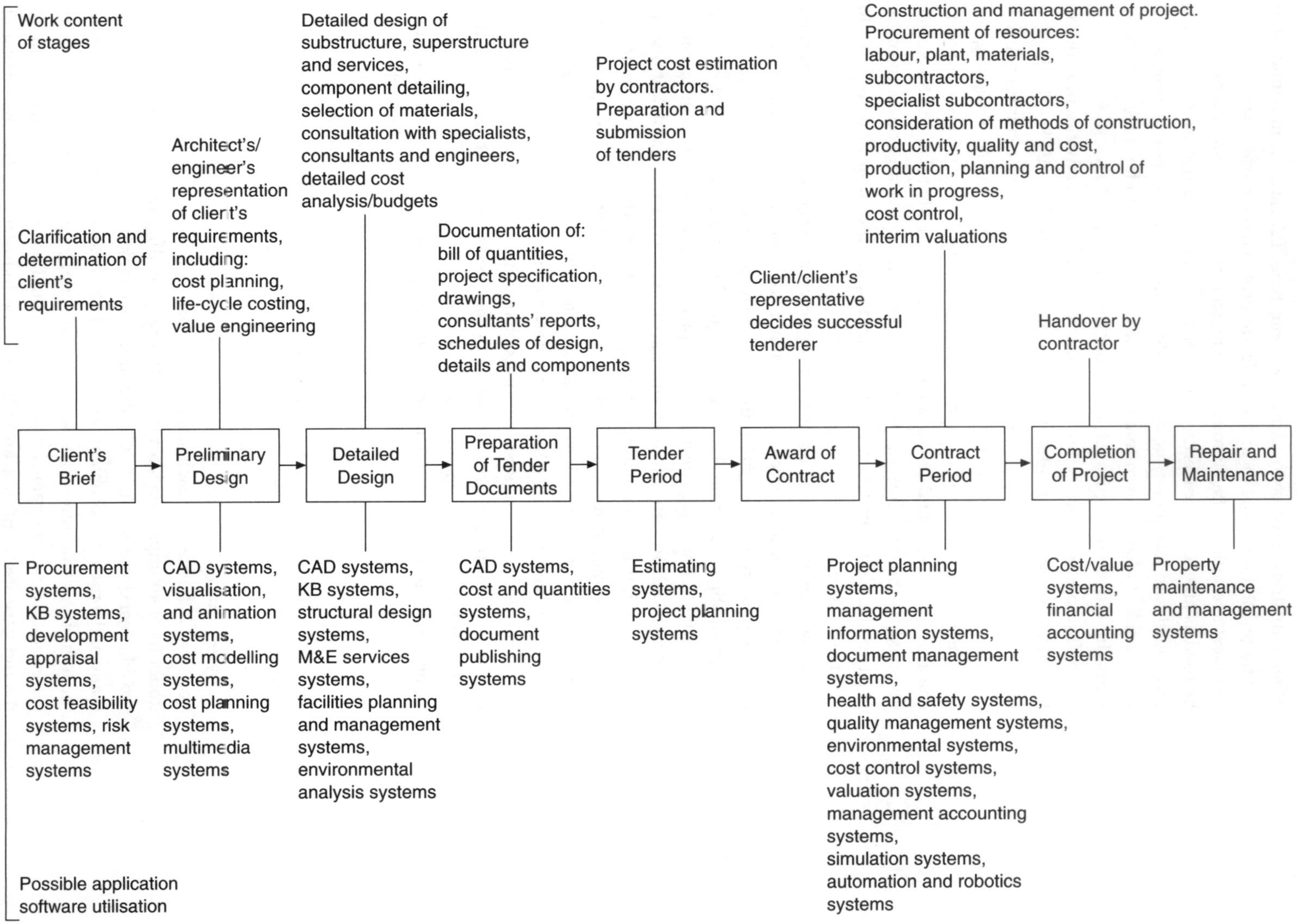

Fig. 7.2 Outline stages of the traditional 'ideal' construction process

having a low capital base, with ease of entry. A contracting firm may start with a small number of employees and become involved in trading activities without computers or reliance on computerised information systems. On a small scale, a company may continue in this way, but as it grows, takes on more staff and larger and more complex construction work, there comes a point where computer use becomes essential. This may be client-driven or forced on a company through required efficient internal modes of working. At some point, a company will eventually become reliant on IT in order to be efficient and competitive. Electronic trading and communications with other parties to a contract may become an essential requirement for future survival.

With large construction organisations, the situation may be somewhat different. Large companies tend to be fully committed to the use of IT and view this strategically as an important part of their business strategy. They will be constantly reviewing their IT infrastructure and have clearly identifiable IT budgets. In-house IT departments may be established to develop and support internal company systems. Such companies will also be keen to embrace new developments, particularly if it can be seen that some competitive advantage can be gained or more efficient and cost-effective modes of working established. The complexity of information management requirements will also dictate the type and sophistication of systems used. Certainly, in other industries downsizing has occurred where companies have moved from large mainframe systems to smaller network systems incorporating client/server architecture. In construction, the reverse has occurred, and this is what many companies are moving up to. From single-user systems in different departments, the desired need for data sharing and information transfer has prompted the use of networking, group work systems and end-user computing. In addition to the applications packages used to support identifiable functions, several software tools are now available that allow users to develop their own applications.

In the past, software was purchased for specific functions. To a large extent, this resulted in the computerisation of existing functions and existing manual systems without linking this to an overall IT strategy. Unfortunately, this can lead to short-term benefits and long-term problems. Without a strategic assessment, situations can arise where an IT migration development path is identifiable, but unachievable, because of the isolated development of individual systems.

The IT developments made to date and in the future will influence the way in which projects are designed, constructed and managed, and these changes will be irreversible. IT provides a range of facilities and communications, resulting in an integrated and productive environment for all parties to a project. Companies which recognise the strategic importance of IT and realise the benefits in the management of information will ensure success in the future.

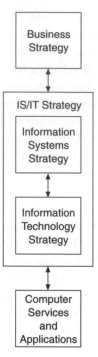

Fig. 7.3 Relationships between strategies

Strategic issues of information technology

The strategic issues surrounding IT are often numerous and complex. When considering strategies, it is important to realise what is involved and what needs to be covered. Various terms are used in describing issues related to IT, so it is often confusing to realise the implications and what is involved (Somogyi, 1993a). Edwards *et al.* (1995) indicate the relationship between various strategies, as shown in Figure 7.3, and consider an IT strategy as the information and systems needs of a business and its related functions. Furthermore, this relates to what a business requires for the future, based on the business itself and its environment.

Clearly, information has strategic importance, and it is often considered an asset without which an organisation would quickly cease to function. Information systems themselves can be manual, partly computerised or totally computerised. It is therefore important to ensure that an organisation's information systems, including those parts which are computerised, are in line with business strategy and anticipated investments are consistent with perceived benefits and risk (Hawley Committee, 1995).

IT strategies in the past, other than in large organisations, have often been based on existing plans. This may have involved bottom-up development, where automation has been introduced related to particular functions. This piecemeal approach often results in isolated technological developments which may not be directly related to the business strategy. Further problems can ensue due to the lack of a strategic approach. System implementation may well prove costly, particularly where priorities change with different information systems requirements. Systems may therefore not integrate, possibly causing duplicated effort, inaccuracy and inconsistency, resulting in poor management of information, and a constraint on the business (Edwards *et al.*, 1995).

While an IT strategy is concerned with information and system needs, this should also include how the needs will be met. This relates to computer services and application development involving the technology and resources required, in addition to their use and control. Certainly, IT is important to organisations in order to improve internal activities, provide required automation, and create and maintain information flows. However, IT should be utilised beyond this to contribute to business administration, involving interactions, communications and co-ordination related to business products and services (Somogyi, 1993a). Thus IT should be used as a catalyst for change 'to informate rather than just automate' (Sprague and McNurlin, 1993).

Moreover, IT itself has important secondary effects on an organisation through manning levels, skill requirements and management systems. This in turn can produce pressure on an organisation, influencing company structure and business relationships. It is therefore important that these issues are addressed in an IT strategy to set direction and identify required management (Somogyi, 1993a).

In addressing these issues, it is important for an organisation to develop an IT strategy appropriate to its needs. This may well take the form of a strategy statement to focus management's attention on business requirements, which will also act as a guide to action and accomplishment (Frenzel, 1996). A comprehensive IT strategy should therefore address the business and technology links within a technology framework, together with details of appropriate delivery and control mechanisms (Somogyi, 1993b), all of which should be supported by senior management and disseminated throughout the organisation.

Furthermore, some form of outline strategy document should be prepared considering the business and its environment which addresses the organisation's goals and objectives. This should also include consideration of an action plan, with assumptions and the risks involved. Additionally, alternative options and dependencies need to be considered, together with resources and financial projections (Frenzel, 1996). Other issues that a comprehensive strategy should address would include the business–technology link; the implications, resulting changes and responsibilities; the technical framework involving systems and applications; and delivery mechanisms and strategy control mechanisms for future review and redevelopment (Somogyi, 1993b). It is therefore important for organisations considering IT investments to first assess their current situation before decisions are made (Construct IT, 1997a).

In terms of technical application characteristics, it is important to separate company requirements from implementation choices. Depending upon a company's trading activities, there may be interactive parts of a system requirement, where some form of network arrangement is required to provide for technical components linking both internally and externally to the organisation (Sprague and McNurlin, 1993; Somogyi, 1993b).

Much emphasis is also placed on IT and business strategy. In addition to the impact internally on an organisation, consideration should also be given to markets, services and products in addition to resources (Edwards *et al.*, 1995). Hillan (1996) describes this issue of integration as a seamless partnership between IT and the business. Furthermore, while IT development may bring about change, impacting on people and an organisation, it also provides the opportunity to exploit IT. It provides greater flexibility in working, and communication opportunities which may well have implications for future markets and work continuity.

While many of these issues can be realised and understood, it is not always easy for companies to decide the best direction in terms of IT developments and implementation and the cost implications. However, there are many well-proven methods and useful strategies to consider and select from in order ensure that best practice is followed (Hillan, 1996). It is also important to realise that IT should be treated as an investment, not purely as expenditure. Failure to move forward and embrace the technologies available to the best advantage may well have an impact on a company's long-term survival.

Costs and benefits of information technology

Investment in IT can often be a complicated process. IT can represent high on-going expenditure, owing to the initial capital expenditure and the time-related costs associated with operating information systems. Certainly, it is important for an organisation to ensure that any investment in automated systems is economically justifiable.

Establishing accurate costs at the outset is not easy. This may well relate to how information systems are procured, operated, upgraded and enhanced, possibly spanning several years. On-going costs in particular, which obviously span the life of a system itself, are difficult to estimate. Readily identifiable costs associated with information systems would include hardware, software, network installations and network software, together with other related costs such as the use of consultants, bespoke software, and the testing and evaluation of new systems (Chalcraft, 1996).

Similarly, benefit evaluation can be problematic, owing to the type of benefits a company can expect. Generally, benefits can be grouped into two categories: tangible benefits and intangible benefits. Tangible benefits are normally readily identifiable and straightforward to evaluate, whereas intangibles can be more difficult to establish. However, with intangibles, it is usually perceived that there will be some end benefit to the organisation or an impact on profitability.

Furthermore, it is important to identify how information systems have an impact on the organisation both internally and externally. Internal improvement may be realised and related to better communications and would also include improvements in workflow, company performance, service to clients, quality and corporate image (Remenyi *et al.*, 1995).

In addition to costs and benefits, however, there may be instances where a proposed information system proves problematic and causes disbenefits. Such instances may include people in an organisation and their reactions where changes have been caused. Additionally, systems may destroy existing practices, and computerised systems may perform well below the level expected.

Information systems expenditure has to be established through formal appraisal of costs and benefits. In establishing financial returns on information systems, several techniques can be employed, including payback assessment, return on investment (ROI), net present value (NPV), internal rate of return (IRR) and sensitivity analysis. Generic weighted methods, where intangible costs and benefits can be quantified, can also be employed. A cost–benefit analysis can also be carried out to compare the costs and benefits of a proposed system implementation.

Establishing a business case for a proposed information system is also an advisable step, supported by documentation. In particular, this should include (Chalcraft, 1996):

- justification for the investment;
- alternative proposals;
- consequences of not proceeding with the proposed systems;
- other business functions and departments affected;
- possible negative effects;
- costs of expenditure;
- benefits of expenditure.

It is important in presenting a case that the investment is seen as positive action, not communicated as a drain on company profits. Moreover, it is important to identify users' expectations and what the system will deliver. The benefits supporting a case for investment will need to be delivered. If this can be assured, future investments are more likely to meet with approval (Chalcraft, 1996).

Analysing and developing information systems

In analysing and developing systems, much will depend on the organisation, its legacy systems, the extent of existing manual and computerised systems and future system requirements. The analysis and design of systems can be carried out in a variety of ways, and several methods have evolved over the years to assist the processes involved.

Methods of analysis and design

Many construction companies have adopted some of these methods in developing systems, and several of the methods are used by analysts and developers in providing bespoke software for construction. Methods of analysis and design include (Yeates *et al.*, 1994):

- structured methods;
- the Yourdon approach;
- Jackson system development (JSD);
- structured systems analysis and design method (SSADM);
- Euromethod;
- Object-oriented analysis and design.

Many of these methods are well documented (Graham, 1994) and are covered in numerous texts. In practice, each method incorporates particular tools and techniques for information collection, documentation and diagrammatic representation. In many cases, the methods also come with automated tools to assist in the processes involved. While no one method can be employed for every problem or organisation, many analysts and designers have adopted a hybrid approach and employed useful features from different methods to develop a required system.

The traditional system development life cycle

In developing systems, many previous and current systems in construction have been developed using the traditional system development life cycle. This represents a basic model of system development and consists of several stages or phases. These include:

- preliminary investigation and feasibility study;
- systems analysis and identification of requirements;
- system design;
- system development;
- implementation;
- operation and maintenance.

The precise naming of the stages and content varies between authors (Hussain and Hussain, 1995), but all follow similar lines in relation to requirements from inception to completion. The stages are shown in Figure 7.4, which also illustrates feedback loops in the life cycle, allowing previous stages to be revised and changes made as appropriate.

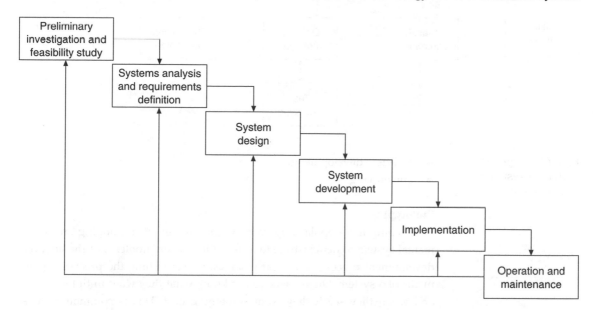

Fig. 7.4 Stages of the systems development life cycle

The preliminary stage is involved with the defining and investigation of particular problems, together with feasibility studies. The systems analysis stage will be a more in-depth analysis of the company in order to understand the current systems in operation and future system requirements. Once established, the new system can be designed, which will involve the interpretation of requirements resulting in a design specification. This will include system interfaces and details of the input and output of data, together with data storage, processing and procedures. The development of the system will follow this stage, where the design specification is used to produce a functional system. This will include the use of software and services for computer programming and testing. Once developed, the working system can be implemented in the workplace and conversion to the new system carried out. Once implemented and operational, a system can be assessed in terms of its usefulness and ability to satisfy user and company requirements.

Alternative development approaches

While the system development life cycle has been used in large construction organisations and for developing large, complex systems, it may not be appropriate for companies with less ambitious system requirements. Alternative approaches can be employed and may be more suitable to specific company requirements. Alternative approaches are well documented (Laudon and Laudon, 1998a; Senn, 1998; Frenzel, 1996; Hussain and Hussain, 1995; Edwards *et al.*, 1995) and include:

- prototyping;
- system development using applications packages;

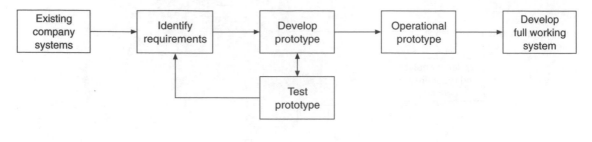

Fig. 7.5 Stages in a prototype systems development

- end-user development;
- outsourcing.

Prototyping

Prototyping is a popular approach which consists of developing an experimental system to demonstrate to a user. One of the problems of the life cycle development process, in the early stages, is establishing the precise requirements of a system. Often users do not know what they want until they see it; and when they see it they want something else. This is common in most industries and is not restricted to construction. Prototyping, therefore, provides a means of quick development and implementation. Precise user comments and requirements can be addressed while a prototype is being built, and iterations in development can take place until a fully developed system results. Figure 7.5 illustrates the stages in the prototyping process.

Applications packages

Another approach to system development is the use of applications software packages. This is probably the most widespread form of system development in construction. There have been considerable developments in applications software for the construction industry, and many packages have been written to cater for a wide variety of functions and processes (CICA, 1998). Many construction organisations have common information requirements and therefore certain packages cater for particular requirements. One major advantage is that packages have been designed and tested and can therefore be relied upon for the processing and functions they perform. Implementing packages can be done quickly, and once learned by users, can produce quick results at identifiable fixed costs. However, there are drawbacks to this approach, since applications packages may not have the required amount of sophistication. It has also been indicated that they satisfy only 70% of organisations' requirements (Laudon and Laudon, 1998a). It is also unlikely that vendors will customise packages to suit individual organisations, which may also be viewed as a disadvantage. Construction organisations in the past have viewed applications packages with the intention and means of introducing computerisation into their organisations. They have often found, however, that company procedures need to fit a package rather than the other way round. In many instances, therefore, construction organisations have decided against purchase and have

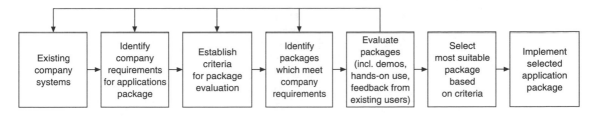

Fig. 7.6 Stages in the evaluation and selection of applications packages

developed their own in-house systems through IT staff or end-user computing. Figure 7.6 shows the approach to the utilisation of applications software packages.

End-user development

End-user development has become popular with many organisation and is likely to become more widespread in the future (Laudon and Laudon, 1998a), certainly in the construction industry. The end-user approach is where users develop their own systems without IT professionals. This is achieved through the use of advanced software tools, which include report generators, screen and graphics design, modelling tools, applications generators and applications languages. As a result, systems can be developed to store and retrieve data, create reports, model work, create web pages and customise information systems to meet a company's needs. There are several benefits to this approach, since users are directly involved in development. Information requirements are clearly understood, and control of the development process is by users themselves.

While this approach is useful for small system development and localised use, it can have implications depending upon the extent of system development. There may be risks due to a lack of appreciation of problem areas and insufficient systems analysis work prior to development. Consideration also need to be made regarding quality standards and the control of systems, particularly where other uses may be dependent upon data and information. End-user development can also produce islands of information within an organisation, and this may be unintentionally hidden from other staff who could benefit from its use. A clear vision needs to be made of the extent of end-user development, since over-ambitious development may have serious consequences for an organisation. Figure 7.7 shows the end-user development approach.

Outsourcing

The outsourcing approach is where part or whole of the information systems and IT services are subcontracted to an external organisation. This may include all the IT infrastructure, telecommunications, network design and management in addition to computing and information systems development.

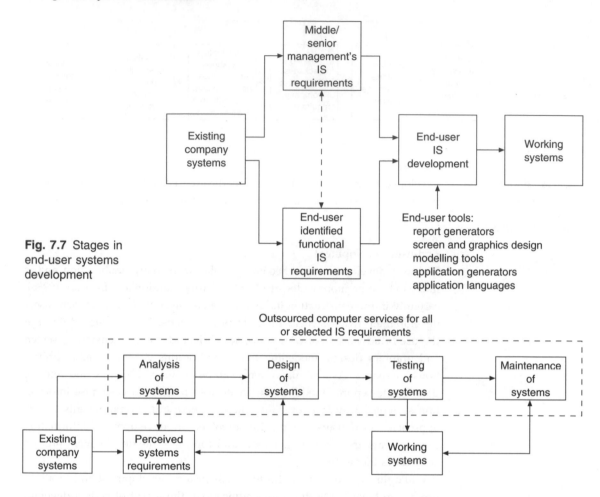

Fig. 7.7 Stages in end-user systems development

Fig. 7.8 Stages in outsourcing systems development

There are several advantages to this approach, particularly where the required resources are not available within the organisation. Buying in the expertise may well be a flexible and cost-effective option, since many construction organisations do not have in-house staff with the necessary IT skills. Outsourcing will provide a balance between resources and the skills required for the successful development of identified and required systems. It also provides the opportunity for an organisation to obtain valuable experience in information systems and their operation.

The disadvantage of outsourcing, however, relates to the lack of control and dependency on the subcontract organisation. There may also be a question of vulnerability with exposure of strategic confidential information. (Laudon and Laudon, 1998a; Frenzel, 1996.) Figure 7.8 illustrates the stages of the outsourcing approach.

System development considerations

In considering the development of company information systems, several issues will have to be assessed before a decision is reached on the approach to adopt. Determining the right system development strategy is not always easy. Situations can prove problematic, and it may be difficult to decide the precise information requirements and the most suitable approach. Newly implemented systems may cause organisational changes, and the impact of these has to be fully assessed. End-user computing, the use of applications software packages or outsourcing may provide short-term solutions but not prove beneficial in the long term. Disparate applications may prove problematic in terms of integration and connectivity. It is therefore important that all approaches are fully evaluated in terms of the long-term impact and the most appropriate selected in accordance with the organisation's information system requirements.

Communications and networks

Information in the construction industry is extensive and varied. The receiving and sending of information is continuous and will depend upon a company's particular involvement in the construction process, and at what stage this is taking place.

The parties to a project may well be representative of the client, architect and designer, together with consultants, statutory bodies, contractors and specialist subcontractors and suppliers. Much will depend upon the project, its size and particular characteristics. Initial information will be generated by the client and professionals involved with design issues, approvals and procurement.

From the contractor's point of view, information will stem directly from clients or their representatives by way of enquiries and invitations to tender. This will be followed by the issuing of tender documents, including the communication of drawings, specifications, bills of quantities, schedules and reports. During the construction process itself, external communications will frequently involve letters of correspondence, architect's instructions, invoices and orders. Additionally, contractors will have internal communications related to production information, resources, costs and income.

While the quality of information may vary, the quantity will, in the main, be continuous from one project to another. Moreover, contractors usually have projects running simultaneously, and during some periods of construction activity, the quantity of information will be considerable.

For information to be useful, issues of timeliness, consistency and accuracy are paramount. For projects to be successful, communication of information is an essential requirement. Furthermore, with demands placed on contractors from the various parties to a contract, communications issues may well have a resulting impact on profitability and survival. However, communications technology is readily available. It is advancing all the time to make the process of communication more effective, efficient and affordable.

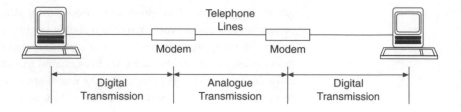

Fig. 7.9 Types of transmission between communicating computer systems

Telecommunications

The use of telecommunications is now widespread throughout the business world. Many construction organisations are now using the technology to communicate electronically and send and receive data and information using several devices. Telecommunications provide the facilities to transmit text, graphics, voice and video. Telecommunications systems comprise both hardware and software to provide for the transmission and receiving of data. In addition to computers for processing, input and output devices are required, together with communications channels and conversion devices, in addition to telecommunications software to support systems and control networks (Laudon and Laudon, 1998b).

The communication channel or medium required can consist of twisted-pair, coaxial or optical cable. Selection will much depend on cost and data-transmission requirements. For high volumes of data over long distances, wireless communication channels offer greater efficiency and may include microwave, radio frequency (RF) or satellite communications (Senn, 1998). The type of signal transmitted can be analogue or digital. Analogue signals are transmitted over telephone lines and are represented by a continuous sine wave. Digital signals, recognised by communicating computers, are represented by on/off electrical pulses, commonly known as bits and represented by a '0' or '1'. Computers can therefore communicate using both signals. The signal during the communication process is converted by the use of a modem from digital to analog, known as *mo*dulation, and then from analog back to digital, which is the process of *dem*odulation (Laudon and Laudon, 1998b; Senn, 1998). Figure 7.9 shows the different types of transmission between communicating computer systems. Additionally, data itself can be transmitted in asynchronous or synchronous mode. Asynchronous transmission involves characters being transmitted one at a time, whereas synchronous transmission involves transmission of data in blocks. Synchronous transmission is therefore much faster in transmitting data and is preferred where it is required to transmit large amounts of data at high speed. Other technical terms used to describe data flow are simplex, half-duplex and full-duplex transmission. In simplex transmission, data flows in one direction only, half-duplex provides for two-way flow of data, and full-duplex allows transmission of data in both directions simultaneously.

Telecommunications technology is a complex area, and the technical details, communicating devices and protocols are well documented in several texts

(Buchanan, 1997). In the business world, and the construction industry in particular, what is important is how the technology can be used. Telecommunications applications and facilities cover several areas, all of which are useful to construction organisations. These include:

- Electronic mail, which provides for the transmission of text messages from one computer to another.
- Voice mail, which captures and digitises spoken messages for storage and transmission.
- Video-conferencing, where both sight and sound are captured and transmitted over a network. This allows a 'live' conference to take place in which participants in different locations can take part.
- Electronic bulletin boards, for posting messages which can be accessed by others.
- Electronic data interchange (EDI), where trading details between organisations can be communicated comprising actual transactions, orders and invoices as opposed to just text.

Much work has been carried out in the area of communications and data exchange, and many initiatives continue to be developed (Hamilton, 1995). Developments have also been achieved by Electronic Data Interchange (Construction) Ltd (EDICON) and the Construction Industry Trading Electronically (CITE) to promote the process of communication. The joining of these two organisations will no doubt help to advance electronic communications in the future.

In terms of telecommunications and the requirements of construction organisations, much will depend on the required form of communications, which will ultimately determine the requirements of a telecommunications infrastructure. Much will be dependent upon the size, location and trading activities of a company. In some situations, centralised computing and communications may be appropriate for a company carrying out work in a limited area, whereas a company carrying out work over a wide area involving regional offices may prefer network communications and distributed processing. Distributed processing can be carried out at regional offices or locations near to where information is required rather than centrally at head office. Computers can therefore be located at different offices connected by an appropriate communications network. In addition to distributed processing, computers can act as retrieval points for data in addition to being sources of data required by others in the organisation. Additionally, companies could use appropriate telecommunications technology and establish a hybrid network to combine both centralised and distributed systems (Senn, 1998).

LANs, MANs and WANs

Network use in organisations can take different forms depending upon precise requirements. Local area networks (LANs) are in common use and are used to connect computers and computer devices over short distances. They

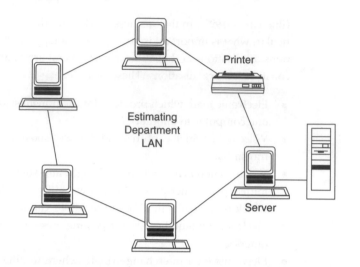

Printer

Estimating
Department
LAN

Server

Fig. 7.10 An
estimating department
local area network

are widely used in offices, office buildings or sites to link together computers
and provide an economical form of computing by allowing hardware and
software to be shared by several users. Figure 7.10 shows a LAN linking
several computers for estimating purposes.

Construction companies requiring the use of e-mail, conferencing and other
on-line applications would therefore find LAN installations useful. Furthermore, companies with several networks may install a network backbone to
provide a common electronic connection to link all networks in an office building. A network backbone provides for high-volume data and allows different
networks to communicate (Buchanan, 1997). Such an installation would prove
useful for a construction organisation with separate departmental LANs requiring connections for data sharing.

Metropolitan area networks (MANs) are larger networks which have evolved
from LAN designs. They provide for data transmission over greater distances
of up to 50 km and operate at greater speeds. They are also able to transmit
more diverse forms of data, including voice, images and video (Senn, 1998).
This form of network would prove useful for construction organisations with
offices and buildings spread over a town or city. Figure 7.11 shows an MAN
layout for a city location where several offices can be linked.

Wide area networks (WANs) are used when data transmission is required
over large geographical areas, possibly covering countries or continents. Communications may therefore consist of switched networks over telephone lines,
dedicated lines or microwave and satellite communications. Organisations
involved in international projects would utilise WANs for data communications and required network applications (Laudon and Laudon, 1998a). An
example of a WAN for a construction company is shown in Figure 7.12.

Communication channels would need consideration when using a particular type of network. For a LAN, an appropriate physical cabling medium
would be selected and installed for a company in its own premises. For MANs

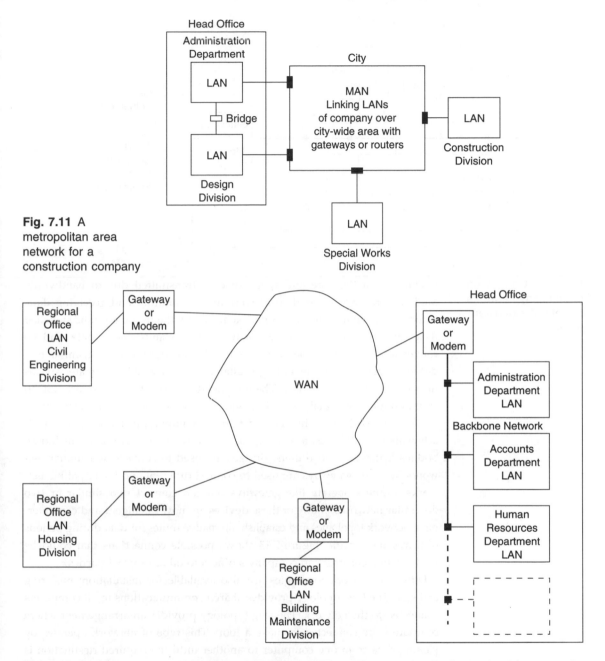

Fig. 7.11 A metropolitan area network for a construction company

Fig. 7.12 A wide area network for a construction company

and WANs, communication channels may take the form of public networks or private networks leased for dedicated use. Value-added networks, which relate to public data communications with additional facilities including storage, and error detection and correction, may also be used.

Communication of data through public networks can also take different forms. A public switched telecommunications network (PSTN) transmits analog data using telephone lines. The network can be used over long distances but has

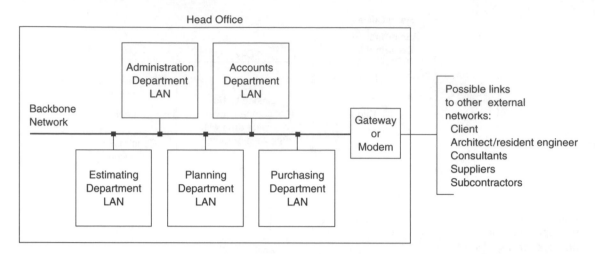

Fig. 7.13 Connection of company networks to networks of external parties

limitations on the type and speed of data transmitted due to bandwidth. Modems are also required to convert for transmitting and receiving data. Direct digital communication can be achieved by the use of a public switched data network (PSDN). This type of network transmits digital data, and no conversion is required when connected to digital equipment. Transmission of different types of data, including digital video, is possible at greater speeds than with switched networks. The integrated services digital network (ISDN) is an example of a digital network for global communications (Buchanan, 1997).

While networks vary in size and type, it is important that connectivity is achievable to allow data sharing. Several devices are available, including bridges, gateways and routers. Bridges are used to connect two similar networks, whereas gateways are used to connect different and incompatible networks. Routers operate like gateways but can connect two similar or two dissimilar networks. Many of these devices are now intelligent and can determine network topologies and establish alternative routes for data during periods of heavy data traffic. Figure 7.13 shows possible connections that could be made from a construction company's offices to other external parties.

Different network topologies are also available for installation: bus, ring and star. The bus topology provides shared communications to all computers connected to the network. The ring topology provides an arrangement where computers are connected to make a loop. This type of network operates by passing data from one computer to another until the required destination is reached. The star topology consists of a central server which switches data around the network. These network topologies are shown in Figure 7.14.

Whichever type of network or network topology is used, the loading on networks can be problematic. Data traffic is not constant but varies during hours of use. It may therefore be useful to assess the volume of traffic at certain times and restrict use of particular applications to avoid network overloading (Buchanan, 1997).

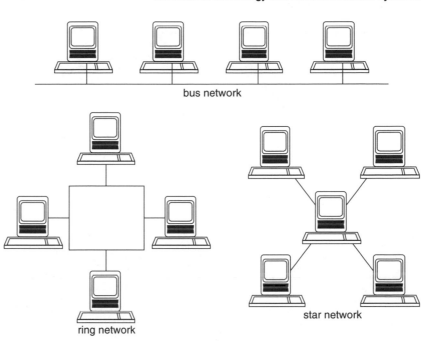

bus network

Fig. 7.14 Network typologies

ring network

star network

Geographically dispersed work is common in construction, and organisations may have several companies or regional offices responsible for particular construction sites and work areas. Communication of data will therefore be required between particular offices and sites. A typical arrangement for a construction company is shown in Figure 7.15.

Network uses in business environments are ever-expanding. It is important for an organisation to assess its data and processing requirements and plan for network use accordingly in order to provide efficient data communications.

Client/server computing

Client/server computing has been around for some time and represents a form of distributed processing. Rather than having separate data created by different applications, client/server systems provide for data sharing through server computing (Gupta, 1996).

A file server holds the data and programs, which are available to users through the interconnection of networks. Both servers and clients are usually connected through an LAN or WAN. Clients request data and information from a server, and this is passed to the client over a network (Senn, 1998). In client/server computing, clients also carry out some processing, the extent of which depends upon the application being used. Usually, this is determined and assigned to the computer best suited to perform the tasks. The operations provided by client/server computing mean that users are unaware that processing may be taking place on different computers. Figure 7.16 illustrates an example of client/server computing.

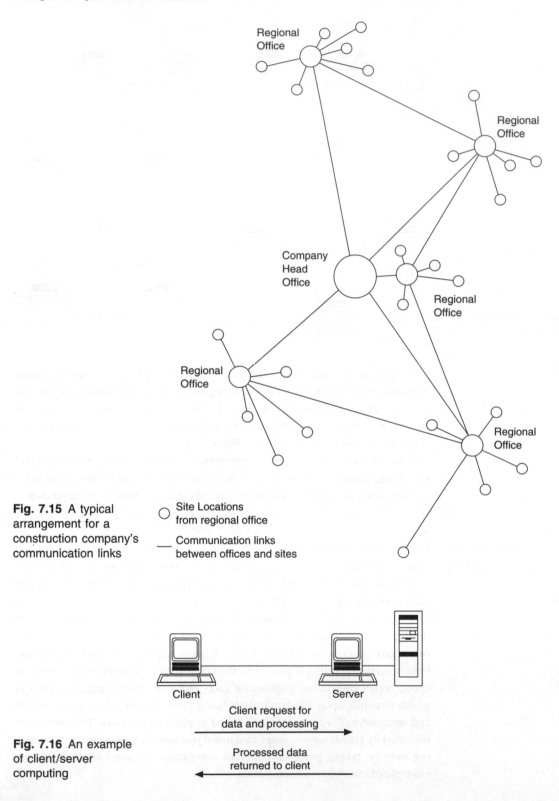

Fig. 7.15 A typical arrangement for a construction company's communication links

○ Site Locations from regional office

— Communication links between offices and sites

Company Head Office

Regional Office

Client

Server

Fig. 7.16 An example of client/server computing

Client request for data and processing

Processed data returned to client

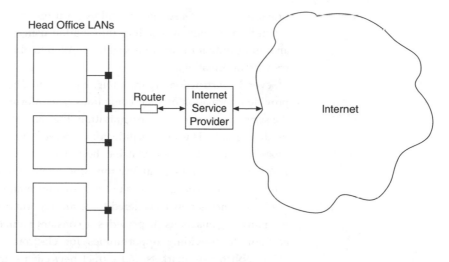

Fig. 7.17 An example
of connecting company
LANs to the Internet

Several benefits are evident from client/server systems. One of the main features is that not all files are transmitted between a client and the server. The only parts transmitted are those relevant to the request and processing. This provides faster access to data and a better service and generally saves time, providing the opportunity for more productive working. Additionally, all users are able to share data and other hardware peripherals connected to a network.

The Internet and the World Wide Web

The Internet, which has experienced considerable growth in recent years, represents an interconnection of networks worldwide. It is also the largest WAN and the most extensive example of client/server computing. Its infrastructure allows the linking of computers located anywhere in the world, providing a vast communications system. This allows individuals and organisations to communicate with each other in the search for and retrieval of information. Figure 7.17 illustrates one method of connecting to the Internet via a router and Internet service provider (ISP). The Internet provides three major services for users: global electronic mailing facilities, information sources and the World Wide Web (Buchanan, 1997).

The World Wide Web (WWW), or Web, represents one of the resources of the Internet and is itself an information source stored on web servers within the network environment. Thus the Web uses the Internet to transmit information around the world. Users access the information using web browsers, these being computer programs designed to access information on the Web. The information on web servers is represented by electronic documents linked via the Internet. These documents can be accessed through the various pages available, which may be text-based in addition to having other multimedia representations including graphics, sound and video. The hypertext pages provide an easy-to-use interface allowing users to point and click, and move around documents in search of information using key words. This provides an opportunity for users to navigate the Web to obtain information as well as provide information.

Several construction organisations have established their own web sites to provide information on their companies. The information displayed relates to the services offered by the organisation, the type of work specialisations and products made. This may include, for example, house construction illustrating types and sizes, the sites available, houses for sale and current prices. Such a web site can be accessed at any time and avoids the need to telephone sales staff or visit the construction site for basic information.

The Internet is now considered to be an important form of communication by many organisations. It provides a constant source of new knowledge in addition to providing opportunities for electronic commerce. Companies can establish new markets and attract new clients, in addition to providing improved customer service at lower cost.

While security and the control of information remain a concern, the number of servers and clients connected to the Internet continues to increase. This results in the expansion of information, and no doubt this expansion will be accompanied by new uses as technology advances. This will also have an impact on how organisations, including construction companies, use future facilities to their advantage.

Intranets and extranets

While the Internet provides communication and other services on a global scale, intranets are private networks where Internet technology is provided inside an organisation. Companies can use their existing network infrastructure for intranets, using Internet standards and web software. Thus intranets can exist alongside established LANs, and legacy systems can be integrated into an intranet (Gralla, 1996).

No special hardware is required, and the software using web browsers provide a common and consistent interface. Web browsers are able to run on any computer, which is particularly useful where companies have different hardware platforms in various locations throughout an organisation. Certainly, this is common in construction, where companies or regional offices have purchased systems or developed software autonomously from other parts of the organisation.

Intranets allow greater collaboration and communication within the company through increased workflow and dissemination. The internal integration and distribution of information related to company processes and procedures can therefore become more effective (Bisset, 1997; Gibb, 1997; Rosen, 1997). Access can also be provided to various forms of company information and on-line documentation. This may be representative of general information, including company news, customer profiles, manuals, appointments and job vacancies, and telephone directories, together with company policy and procedures, health and safety, and environmental issues. Communicating such information company-wide eliminates the need for paper copies, their printing and distribution (Hinrichs, 1997; Tod, 1997).

Individual departments in the organisation can develop their own web sites to support and inform other employees in the organisation. For example, in a construction organisation, the estimating department could display details of tenders currently being prepared, together with successful tenders and value. The planning department could have details of projects and current progress and completions. The plant department could provide information on items of plant, their current location and availability. The human resources department could also provide details of trades and operative availability, job posting, training courses and opportunities. Thus in a large geographically spread organisation, fast, low-cost communication of information would bring people closer together by providing the quick retrieval of relevant information, facilitating increased workflow.

The extent of the availability of company information would be a decision for senior management. The ease of access of information through Internet/intranet technology may well restrict its use for sensitive company information. Organisations may decided to restrict such information, particularly financial and issues related to human resources management. Such information would be best separated and accessed through applications running on LANs and WANs. Thus companies may decide to operate both types of technology in a complementary way, providing a hybrid environment (Coleman, 1997). Intranets therefore provide a useful infrastructure for collaborative working within an organisation and may well be considered an essential part of an organisation's IT portfolio.

As an extension of intranets, extranets can also be used, and these can be described as private intranets. They provide accessibility by selected outsiders to the organisation. Thus they are useful for linking an organisation with its clients, subcontractors or suppliers, and provide cost savings through more efficient communications.

Groupware and group support systems

Groupware can be described as software, and also hardware, that is used to promote collaborative working. It supports team working for people who need to work together on a particular project. For example, in addition to electronic mail and messaging, people may require to work on a single document, carry out group scheduling or require real-time exchange of data or video-conferencing. Such facilities therefore encourage collaborative working and provide better co-ordination and integration of geographically located staff (Simon and Marion, 1996). However, for groupware to be successful in its implementation, a company would need to consider its processes and organisational culture as much as the technology. Without this, it is unlikely that groupware would reap the expected benefits (Coleman, 1997). It would therefore be important for an organisation to consider its own requirements and the particular choices for sharing, collaborating and distributing information.

Essentially, organisations have three options for communicating: the Internet, establishing a company intranet or using groupware. In selecting an option, much will depend on particular requirements (Buchanan, 1997). Arguably, the Internet and intranets provide for collaborative working and are continuing to advance in the sophistication of features available. However, the Internet protocol may not support the precise requirements of users and may not provide the functionality and reliability of LAN-based groupware (Coleman, 1997).

Certainly, for construction organisations, groupware can prove beneficial for collaborative working and has been used on projects for communication between clients, architects and contractors in order to distribute and share essential project details (Black, 1997).

This concept of group working is also provided by the use of group support systems (GSS). An essential feature of these systems is on-line interaction. Group participants can ask questions, and communicate ideas and suggestions for comment and feedback. Such communications can be captured electronically in a database and provide information for decision making on specific problems (Senn, 1998).

Such systems could therefore be used on projects where problems occur on site, and where decisions have to be made related to a practical situation. Group support could therefore provide the opportunity to discuss, for example, a problematic construction operation requiring immediate attention, where construction methods, resource requirements and cost consequences have to be assessed and a speedy decision reached. Allowing correct decisions to be made before proceeding may well avoid serious implications related to other operations, subcontract work and project completion time. Without the appropriate communications technology, such occurrences may prove difficult to avoid.

Virtual private networks

Virtual private networks have gained in popularity for connecting various offices in organisations. These networks allow remote users to be connected via the Internet, the main advantage being the low cost involved in the use of public networks and avoiding the use of permanent or leased lines. Moreover, obtaining this form of networking by outsourcing may improve service and reliability, particularly for remote users, in addition to overcoming problems of scalability and manageability (Green, 1998).

While this form of networking does provide cost-effective communications, security becomes a major concern. Company information will be communicated over public networks, and sensitive information will need to be secure. Providing that encryption and the authentication of users can be established, these networks have immediate benefits related to issues of mobility.

Construction organisations, in particular, would find such technology advantageous with geographically dispersed construction projects requiring communication links with company offices and other external parties.

Network privacy and security

While the communication of information through networking and the use of telecommunications provides numerous benefits, network privacy and security are issues that need serious consideration before design and implementation. In terms of IT, privacy of data relates to personal or organisational information and, in particular, how such information is collected, stored, used and protected. For network installations, this has important consequences for control and access. Individuals would not welcome personal details being misused, just as organisations would not want sensitive information to fall into the hands of competitors.

While legislation may indicate the requirements of privacy, organisations, and particularly those in control of information, should follow ethical standards in terms of IT use in a business environment (Hussain and Hussain, 1995). Networks in particular are vulnerable to intrusion. Sensitive organisational information should have restricted access from internal personnel as well as intrusion externally (Gralla, 1996). An intranet designed to allow users access to the Internet also creates a situation that allows intruders to enter the intranet.

The issue is one of deciding the type of security required. Admittedly, incorporating security measures in a network can reduce performance, but external connections to networks can also provide the opportunity for intrusion. Company data, and business correspondence generally, are often confidential and should therefore be treated as private. One major problem is that in using telecommunications systems, data passes through numerous transfer points where individuals may have access to intercept and read messages (Coleman, 1997).

Certainly, the networks used by companies have vulnerabilities. Much is due to the technology used, the different modes of entry to networks and the way in which data are exchanged. This is particularly so where companies have a comprehensive network arrangement within a distributed environment. Coupled with this is the amount of data traffic on the networks, which is often difficult to track in relation to where it may be any one time (Galla, 1996). Security measures are therefore needed, and much will depend upon the networks in an organisation, the links externally, and the value and confidentiality of data being transmitted.

In providing security of systems, several approaches can be adopted. Firewalls, which are hardware and software combinations enabling users of an intranet to access external networks, also prevent unauthorised users accessing the intranet of the company. This is illustrated in Figure 7.18. Additionally, routers can play a role in a security system, providing a filtering facility

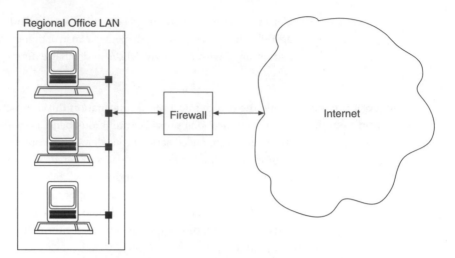

Fig. 7.18 A representation of a firewall between networks

by examining packets of data before allowing or preventing access (Gralla, 1996). Additional security measures may also be employed through encryption of a message using a conventional scheme based on encryption data standards. Authentication of messages may also be a requirement based on digital signature security and encrypted with a private key (Coleman, 1997).

Internal security measures may also be desirable in an organisation. In addition to encrypted intranets, passwords are commonly used to provide authorised access, and classified areas may be protected and access restricted to certain members of staff. Companies may also decide to establish physical privacy and security through the use of different servers and network arrangements for sensitive and business-critical data (Hinrichs, 1997).

As part of security, companies should also establish some form of disaster recovery plan. This is essential where companies are dependent upon their computing and network facilities. This should detail how a company is to maintain its information systems and resume operations based on identifiable methods and actions. This may entail establishing a back-up site and access to other computing and telecommunications equipment. Such facilities may be provided by the use of a bureau or through some previously agreed outsourcing arrangement (Gupta, 1996). Certainly, it would be essential to identify the business-critical data and operations and to know the computing and human resources required to resume operations as quickly as possible. It is an organisation's responsibility to assess the risks involved and determine how systems are to be protected. Security measures must also be clearly communicated to users.

Security itself is a dynamic process owing to the rapidly changing environment and technological advances in computing and telecommunications. Companies have to decide what measures need to be taken, and it is important to establish the right balance between security and the performance of information systems.

Implementing information systems in construction organisations

The implementation of information systems is concerned with all the processes involved in installing a new system, including its operation. This may represent the conversion from an old system to a new one, or the possible installation of a computerised system to replace existing paper-based systems. In doing so, this will in some way cause changes to the business operations or procedures and possibly affect the flow of information through the organisation. As with other projects, such a conversion will need to be planned before the benefits of the system can be realised, because failing to do so may result in problems concerning time, cost, and attitudes towards the system and its acceptance. Much preparatory work, which essentially revolves around equipment, particularly hardware and software, will have to be initiated.

Hardware and software selection

The selection of hardware and software will have to be made at some stage and will depend upon the particular approach used in developing an information system. For systems that have been developed following a development life cycle, early selection will be made during the appropriate stages of development, and particularly when systems are being designed. For systems that are developed using applications software, this will relate to the particular requirements of the package used.

For hardware, upgrading may be required or the purchase of new computers. In terms of selection, requirements will be based upon size and capacity, memory, speed of processing, and communication features. Comparisons will need to be made between computers, their particular features and performance. Additionally, financial considerations need to be made. This relates to whether equipment is to be purchased, rented or leased. The rent and lease options are probably more expensive but do have the advantages of flexibility and a reduced likelihood of obsolescence. Purchasing equipment will call for cash payments, and ownership will also bring the responsibility of dealing with equipment problems. Maintenance and support therefore become important requirements, and considerations need to be made in terms and service, in addition to response time and cost (Senn, 1998).

Software selection is equally important. The evaluation of software should address issues of reliability and capacity, and the maintenance and support available from a vendor. Chosen software should be flexible in order to deal with changing requirements and the needs of users. When system requirements are known, commercial software may be appropriate and an identified package may meet requirements. Comparisons with similar software are advisable, and criteria should be established to consider user-friendliness, file characteristics, ease of use, hardware resources, user documentation, maintenance and cost.

Other equipment and resources also need to be considered prior to implementation. This relates to the location of the system, whether it is to be office-based or temporarily on site. The extent of such preparation work will also depend upon the system and the particular circumstances. The work may be minimal and involve the replacement of computers for new ones to support the new system, or it may be more substantial. A large system may require new and additional equipment and furniture. Additionally, air-conditioning, lighting and security systems may be required. If implementation involves several computers based on an LAN installation, cabling, connections and outlets will have to be installed to accommodate requirements (Senn, 1998).

It is therefore important that such work be carefully planned. A schedule of activities should be prepared with identified responsibilities, and the work should be carefully monitored and controlled. In addition to the equipment, consideration should be given to file conversion, and particular data formats and data requirements, including documents and controls. People in the organisation should also be made aware of the changes, since this may have a direct or indirect effect on their own working practices and procedures.

Implementation methods

Several implementation methods are available (Senn, 1988):

- direct changeover;
- parallel systems;
- pilot systems;
- phased-in systems.

Direct changeover

The direct changeover method is an immediate conversion to the new system. This will be appropriate where time is a governing factor and where no dramatic changes are occurring. This method has the advantage of forcing people into using the new systems but may be risky and is rarely problem-free. However, because there are no systems to go back to, this may force people into being more committed to making the implementation successful.

Parallel systems

The parallel system approach is a more cautious one that allows both old and new systems to run together in parallel before the old system is abandoned. This method does provide the opportunity to fall back on the old system if there is some uncertainty with the new system, but it may prove problematic in convincing people to convert.

Pilot systems

The pilot system method is useful where major changes are being introduced. A small working version of the system is introduced to a selected group or particular department. This provides the opportunity to learn from the pilot system and solve any problems experienced. When the pilot system is working successfully, implementation can be piloted elsewhere until the entire system is operational.

Phased-in systems

The phased-in system approach has carefully planned phases of implementation. It may not be possible to install the system in its entirety, and prioritisation may be appropriate to install hardware and software and operate the system before another phase is introduced. This also allows problems to be isolated and dealt with. This does allow some employees to use the system before others, and some clear phase-in plan should be communicated throughout the organisation. The phase-in should be completed as quickly as is practicably possible to allow all eventual users to gain the benefits of the system.

Human–computer considerations

The human element of information systems is one which is often overlooked, or at least underestimated, yet acceptance by users of a system is paramount to its success. Resistance to computers and their use is common in organisations, because computers often signal change, and this can cause concerns among employees. People are often disturbed by change, particularly with computer technology, since this can often affect their sense of control and purpose (Hussain and Hussain, 1995).

The change of procedures resulting from the introduction of computer systems can cause restructuring within an organisation and alter people's positions and status. Additionally, change can influence people's perceptions of employment, security, authority and interaction with other employees. The resistance to change of a new computer system may well be relate to the computer being unable to relate to specific job requirements, mistrust of the machine, possible loss of data, concern about user mistakes in using the system, and health concerns over use. Likewise, managers may have concerns about computers influencing their management role and affecting the decision-making process (Yeates, 1994).

The human issues surrounding the use of computers need to be addressed at an early stage. Systems which are user-friendly usually meet less resistance. A system that is easy to follow and sufficiently intuitive should meet with approval. Certainly, human–computer interaction issues should be considered at the design stage of system development, or during software evaluation and selection. Users should also feel comfortable with a system, have a sense of control, and be able to evaluate and correct their stored input data. The system itself should also be sufficiently adaptable to suit users from different backgrounds and levels of proficiency (Hussain and Hussain, 1995).

Overcoming resistance to change and the adoption of systems is a management issue. This is invariably concerned with behaviour and attitudes towards a system and a knowledge of human–computer interaction, and communication skills are required to convince people of the benefits. People should be given the opportunity to participate, and to evaluate systems, and be given the time to learn the skills required for operation. Once they have seen and experienced the worthwhile benefits, they will be more ready to accept and use a system.

Staff education and training

The education and training of staff in connection with information systems is of paramount importance. If perceived benefits – increased productivity levels and efficiency gains – are to be realised, measures need to be taken to provide appropriate education and training. The way in which systems are operated and used needs to be reinforced and will ultimately affect the success of system implementation.

Education and training can be carried out at various levels, depending upon the size and complexity of the system. General education programmes informing personnel of the overall information system structure and how it is to operate need to be initiated. This will give some overall understanding of systems in place, and how company information is to be processed and communicated throughout the organisation.

The extent of training requirements will vary and be dependent upon the organisation's information systems infrastructure. Companies with complex systems involving the processing and communication of information between regional offices through networks and distributed systems will need information system managers, system operators and support staff. Such staff will need training in the operation of equipment and support services in order to deal with problems as and when they occur.

In small organisations, where information systems are represented by departmental computers and software, personnel may well have the responsibility of operating and managing computer systems in addition to being users. This may be where company systems have been developed by end-users or may be through the use of several applications packages. Whatever the situation, personnel will have to be trained in operating equipment and software for processing and reporting, and be able to deal with problems as they occur.

Various options for training are now available, and organisations need to assess their training requirements in relation to their information systems. General training courses are provided through software houses and universities, and much will depend upon company requirements and the location of training centres.

With regard to specific training related to particular products, training from the supplier is probably the best available. Many companies which supply software have training divisions whose main purpose is to support customers. Training can usually be provided at the supplier's premises, involving hands-on sessions with trainers who have an in-depth knowledge of the software product. Scheduled courses may also be advertised relating to training on specific products. Additionally, in-house training may be available from the supplier, where training can be tailored to suit the organisation. Several personnel can therefore be trained, although this may be dependent upon the organisation's own facilities available. Hands-on sessions may be limited, and work-related interruptions may also prove a disadvantage.

The training supplied should also include appropriate training materials in the form of tutorial guides and workbooks, with examples to follow in order to consolidate the learning experience. Certain learning objectives should be communicated by trainers, and trainees should be confident that they have acquired the skills necessary to use a system. Further materials may be available in the form of user guides and reference guides to use in the workplace during operation of the systems.

Education and training of personnel in the operation and use of information systems is an integral part of development, implementation and operation. It is essential that appropriate training take place and be provided for operators and users of company systems. The costs involved should therefore be viewed as an investment in the company in order to achieve the expected benefits.

Effective information systems management

Following implementation and operation, management of information systems becomes an on-going commitment. The information systems and facilities, together with company data, information, and the people operating and using the systems, represent an important company asset, which needs to be managed accordingly (Beare, 1997; Hawley Committee, 1995).

Post-implementation review

A useful starting point in the management process is a post-implementation review. This should provide information on how the system is working in terms of meeting its objectives, operational efficiency and acceptance by users. Many factors influence the success of information systems (Whyte and Bytheway, 1995), and it is important to evaluate the issues related to efficiency and effectiveness.

Information can be collected from personnel by questionnaires, interviews and observations. Common concerns of users are normally associated with ease of use, user confidence, identified benefits and output quality. However,

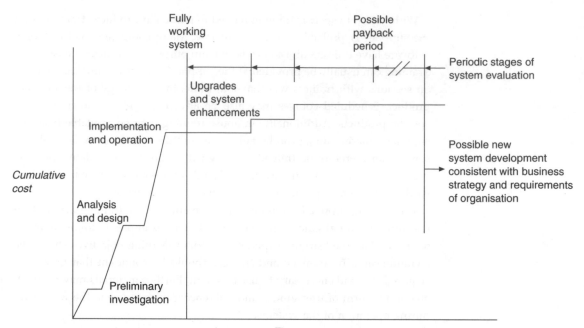

Fig. 7.19 Cumulative costs associated with an information system

other operational issues will also be of interest concerning timeliness of operations, validity and completeness of reports, frequency of errors, response time, and reliability (Hussain and Hussain, 1995).

Such information will be useful not only for maintenance purposes but also for future internal development to support new business initiatives. Evaluating information systems should not therefore be a one-off exercise but should represent a continuous participatory process throughout the life of a system (Remenyi *et al.*, 1997). Figure 7.19 shows the cumulative costs associated with an information system, illustrating periodic stages of system evaluation.

Management considerations

Many issues will arise during the life of a system, and appropriate organisational management will have to be introduced accordingly (Edwards *et al.*, 1995). In particular, management considerations will include:

- documentation;
- use of systems;
- security;
- system support.

Documentation

Documentation should cover system documentation, operation documentation and user documentation. Each serves a different purpose but is equally important. System documentation is particularly important where systems have been developed within the company or are the result of end-user computing. Details should be provided of how the system has been developed to assist in dealing with future enhancements. Technical manuals should also be available on other products used in the development process. Operation and user documentation may be produced internally within the organisation. Operation documents provide details of how to use equipment, and user documentation provides guidance on using software. Where applications packages are being used such documentation should be provided by the supplier but may also be incorporated in the software. Documents produced internally could also be made available electronically, through intranets if applicable.

Use of systems

Correct use of systems by employees is also a management concern. Company documentation should clearly state how systems are to be used, indicating company policy, if appropriate, which should promote good practice for the input and processing of data. This is particularly relevant where systems have internal links to other part of the organisation, and external links to other parties.

Security

The security of information systems covers a variety of issues but predominantly access to software and data, which is particularly relevant where remote access to computers and networks is possible. Other issues relate to physical damage and theft, and adequate protection arrangements need to be made accordingly. Data back-up and recovery procedures need to be assessed, with appropriate methods being employed in case of system failure. Special arrangements also need to be made for the use of compatible equipment or the use of contracted facilities.

System support

System support will depend much on system complexity and use. Suppliers should include support by way of hot-line communications for queries on software use, with call-out support for hardware failures and major software malfunctions. Annual maintenance contracts, which may also be inclusive of periodic software upgrades, can be arranged to cover such eventualities.

Future developments

The speed of technological developments invariably seems to accelerate. The reality of technology often surpasses what was previously perceived. Information systems through advances in software development environments will no doubt become more powerful and flexible, influencing data capture, storage, retrieval and communication.

This is likely to have an impact on the construction process, and automated systems with intelligent components will provide a continuum of data flow and information generation throughout the various stages of a project. Future priority challenges for the construction industry have been identified as integration of information and processes (Construct IT, 1997b). This will no doubt be achieved through complex information systems that will be linked to construction processes on site. Intelligent automated operations and robotics may become widespread as an integral part of construction. The technology will also have to embrace, and be sensitive to, human issues, but the power and flexibility of future information systems may change the way we think, construct and manage projects.

The communication of construction will also take on new dimensions with advancements in telecommunications. This will not only influence internal company communications but also external links with other parties to a project. Powerful wireless networks and mobile data networks may be a requirement for construction organisations, where remote information access to and from offices and sites becomes essential. Furthermore, networks of the future will have to be sufficiently flexible to detect messages for prioritisation where information is required to assist in the decision-making process. The sophistication of communications will also be a requirement for international contracting organisations operating on an global scale. Autonomously developed company systems will need to communicate and provide seamless and cohesive information systems throughout an organisation.

It is also likely that technological advances in information systems will provide numerous opportunities for construction organisations. Companies that are prepared to invest in and embrace such technology and systems will be well equipped to operate in future markets.

Summary

This chapter has been concerned with information systems and the communication of information. Information that is generated during the construction process comes in different forms, from different parties to a project, and is often extensive. To cope with this information during construction, IT is now playing an important role, and numerous software applications are available for different functions and activities. These have been illustrated in relation to the stages of the construction process.

IT and its use in organisations has strategic importance and needs to be considered in line with business strategy. The costs and benefits of IT also need to be addressed in support of information systems, since these will impact on an organisation's working operations. Investments in IT can involve considerable capital outlay, therefore the correct approach needs to be adopted in the analysis and development of information systems. The different methods of development have to be considered, from the traditional system development life cycle for large systems to the use of applications packages to support specific functions. Development considerations are therefore important in order to decide on the most appropriate approach for an organisation.

Coupled with information systems is the issue of communication. This applies internally within an organisation in addition to external communications with other parties. The communication of information by the use of telecommunications is becoming commonplace. Networks have therefore been considered from local to wide area networks for the communication of data. This was further extended to cover the Internet, the concept of client/server computing and the World Wide Web.

Attention has also been given to Internet technology within organisations through the use of intranets and extranets. Furthermore, groupware and group support systems also provide the opportunity for greater efficiency in workflow and group working. Virtual private networks are also useful as part of remote-access communications. However, these networks, as with other networks, need careful attention with regard to privacy and security. These issues are extremely important and need to be addressed as part of network design and implementation.

The implementation of information systems generally in organisations involves several activities, and appropriate implementation plans need to be prepared and monitored accordingly. Hardware and software need to be assessed in line with developed systems and the requirements of the organisation. Implementation methods have been considered, and covered in more detail, since they can be influential in the success of new systems. Important points relating to implementation are concerned with human issues and also the education and training of staff. Various ways of obtaining training have been discussed, and it is important to ensure this is carried out as appropriate.

Information systems management has also been addressed. An important starting point is the post-implementation review, and ideally the review process should be an on-going commitment for an organisation.

Future developments in information systems are taking place all the time, and while it is difficult to predict specific advances, it is likely that systems will become more sophisticated and intelligent. Communication systems will also become more widespread and, in particular, will increase in efficiency and accessibility.

References

Beare, D. (1997) *Effective Information Systems*, Pitman Publishing, London.

Bisset, M. (1997) 'Time to get wise to the use of intranets', *Computer Weekly*, 2 October, p. 34.

Black, G. (1997) 'Service to build faster links for construction industry', *Computer Weekly*, 9 October, p. 14.

Buchanan, W. (1997) *Advanced Data Communications and Networks*, Chapman & Hall, London.

Chalcraft, C. (1996) *The Justification and Costing of Information Systems*, Technical Communications (Publishing) Ltd, Hitchen.

Chartered Institute of Building (CIOB), *Construction Computing*, quarterly publication, Ascot, UK.

Coleman, D. (1997) *Groupware: Collaborative Strategies for Corporate LANs and Intranets*, Prentice Hall, New Jersey.

Construction Industry Computing Association (CICA) (1998) *Software Directory*, annual publication, Cambridge, UK.

Construct IT (1997a) *A Health Check of the Strategic Exploitation of IT*.

Construct IT (1997b) *Research Futures: Academic Responses to Industry Challenges*.

Edwards, C., *et al.* (1995) *The Essence of Information Systems*, second edition, Prentice Hall, London.

Frenzel, C. W. (1996) *Management of Information Technology*, Boyd & Fraser, Danvers, Mass.

Gibb, S. (1997) 'Getting wired: Internet, intranet and extranets – What is the net benefit?' *Project*, Vol. 10 (5), p. 26.

Graham, I. (1994) *Object Oriented Methods*, second edition, Addison Wesley, Wokingham.

Gralla, P. (1996) *How Intranets Work*, Ziff-Davis Press, New York.

Green, T. (1998) 'Accessing all areas', *Computer Weekly*, 2 July, p. 41.

Gupta, U. G. (1996) *Management Information Systems: A Managerial Perspective*, West Publishing, Minneapolis.

Hamilton, I. (1995) *EDI STEP AND CALS*, Technical Paper No. 3, CICA, Cambridge, UK.

Hawley Committee (1995) *Information as an Asset*, KPMG, London.

Hillan, I. F. (1996) *Integrating IT and the Business*, Technical Communications (Publishing), Hitchen, UK.

Hinrichs, R. J. (1997) *Intranets: What's the Bottom Line?* Sun Microsystems Press, New Jersey.

Hussain, K. M. and Hussain, D. S. (1995) *Information Systems for Business*, second edition, Prentice Hall, London.

Laudon, K. C. and Laudon, J. P. (1998a) *Management Information Systems: New Approaches to Organization and Technology*, fifth edition, Prentice Hall, London.

Laudon, K. C. and Laudon, J. P. (1998b) *Information Systems and the Internet: A Problem Solving Approach*, fourth edition, Dryden Press, Fort Worth, Texas.

Nunn, D. (1997) 'Step into the future', *Contract Journal*, 12 March, pp. 20–1.

Paulson, B. C., Jr (1995) *Computer Applications in Construction*, McGraw-Hill, New York.

Remenyi, D., *et al.* (1995) *Effective Measurement & Management of IT Costs & Benefits*, Butterworth-Heinemann, Oxford.

Remenyi, D., *et al.* (1997) *Achieving Maximum Value From Information Systems: A Process Approach*, John Wiley & Sons, New York.

Rosen, A. (1997) *Looking into Intranets & the Internet: Advice for Managers*, American Management Association, New York.

Senn, J. A. (1998) *Information Technology in Business: Principles, Practices and Opportunities*, second edition, Prentice Hall, New Jersey.

Simon, A. R. and Marion, W. (1996) *Workgroup Computing: Workflow, Groupware, and Messaging*, McGraw-Hill, New York.

Somogyi, E. (1993a) 'What is the purpose of IT strategy?' in *Information Systems Practice: The Complete Guide*, Tony Gunton (ed.) NCC Blackwell, pp. 5–12, Oxford.

Somogyi, E. (1993b) 'What issues should a comprehensive IT strategy address?' in *Information Systems Practice: The Complete Guide*, Tony Gunton (ed.) NCC Blackwell, pp. 13–22, Oxford.

Sprague, R. H., Jr and McNurlin, B. C. (1993) *Information Systems Management in Practice*, third edition, Prentice Hall, New Jersey.

Tod, E. (1997) 'What users really, really, want from an intranet', *Computer Weekly*, 6 November, p. 38.

Whyte, G. and Bytheway, A. (1995) *Factors Affecting Information Systems Success*, Cranfield School of Management, Cranfield University, Bedford, UK.

Yeates, D., *et al.* (1994) *Systems Analysis and Design*, Pitman Publishing, London.

8 The influence of management systems on organisation and human resources

Introduction

There has been a profound change in the orientation and practice of management by organisations over the last 20 years. The customer-focused marketplace and fierce competitive service positioning have demanded clear attention to performance improvement and added value delivery. Changing and more demanding business environments have highlighted the need for dedicated strategic consideration coupled with efficient and effective directive and operational management. Organisational change appears to have become synonymous with emotive exercises in outsourcing, downsizing and the ubiquitous application of re-engineering. Such practices have been widespread in recent years, and their effects have often been radical, severe and not without resistance and dysfuntionality in many organisations.

In recent times, there has been a considerable culture shift from the morphostatic organisation, with its traditional and often outdated business practices, to the morphogenic organisation, in which change is proactively championed and even revered as being a prerequisite to business dynamicism (Griffith and Watson, 1998). Such change has significantly influenced employment structures and the skills and attributes valued by organisations (Anderson and Marshall, 1996). In the 1970s, the wholly bureaucratic structures and strictly prescribed tasks had valued in employees the educational basics such as literacy, honesty and reliability. In the 1980s, a greater focus on business operations and the drive for quality led to a preoccupation with personal skills development such as communication, self-desire and assertiveness. In the 1990s, the flatter lean-organisation culture, where more employees are responsible directly for parts of the business, has led to a greater need for much broader commercial vision and the targeted delivery of outcomes by empowered, project-management teams.

With the evolution of many organisations into the morphogenic type, external influences have become as important as intra-organisational aspects to structure and human resourcing. Greater cohesiveness in management is needed to accommodate more demanding customer requirements in the marketplace. Competitiveness must be understood better to ensure the added value of products or services. Increasingly stringent legislation, for example in environmental matters and employee health and safety, must be recognised and responded to. Such important factors require that organisations are clear in the designation of management roles and that managers are supported in their enabling and empowering functions.

Human resources management has taken over from traditional personnel management. It advocates a holistic approach where personnel matters are given greater attention by line managers and are considered with direct reference to the core business planning of the organisation. Emphasis is placed on the business objectives and on relating these to the performance management of teams and individuals. Bringing the efforts of the different business team strands together forms the basis of organisational systems management.

In many organisations, the structure and organisation of a management system and the parts, or sub-systems, that make up the system are relatively simple. This is because many businesses are based on a single and central corporate management location with single, or a small number of, production sites. Construction organisations differ from this in well-recognised and accepted ways, for example in their geographically dispersed decentralised management, temporary organisation, multi-project situation and interdisciplinary elements. Such factors often dictate that different, although interrelated, systems must be established for corporate management and for the management of each project organisation. In effect, two tiers of management need to be established.

In construction, the effective management of both the corporate organisation and the various project organisations is dependent upon the synthesis of many interdisciplinary elements and resources. It is the functional interaction of all the parts producing the necessary synergistic effects that leads to the successful delivery of a construction project and hence contributes to the well-being of the corporate organisation.

Management systems

Systems management concepts have permeated management thinking for well over two centuries. To some extent, a systems approach can be advocated in any organisation as it meets a number of the most basic organisational needs.

A management system is, simply put, a way of doing things. Systems develop protocols and sets of procedures which bring structure, order and therefore stability to an organisation where otherwise there might be a potential for chaos. If configured as a holistic and morphogenic open system, i.e. one which interacts with its surroundings, it can meet the wider organisational needs.

To meet the basic need for organisational existence and development, any management system must itself meet several fundamental requirements. A system should be simple but not simplistic; i.e. it should be easy to understand, interpret and implement by the people who work with and around it. A system should give reliable and consistent outcomes. A system must be capable of being translated into easily conducted sets of procedures and tasks. This is particularly important in construction, since at management level it is function-oriented and at project site level it is task-based.

In many organisations, the term 'system' is frequently misunderstood. It is not uncommon for 'management' to be confused with a 'system'. The system is really the organisation that exists to serve the core business. Management is the control of the sub-systems, usually structured into functions around specific management concepts, for example quality management or health and safety management. These specific concepts are brought together through their management approach to support the core business of the organisation.

Figure 8.1 illustrates the relationship of various organisational sub-systems to an organisation's system. It can be seen that the core business of the organisation is affected by a great many internal and external influences. These influences determine the main parameters for the existence and operation of the organisation. The core business can be serviced and supported by as many sub-systems as are needed. In the illustration, only the management systems forming the substance of this book are shown. Each sub-system is essentially a specialised and dedicated management function, for example environmental management, financial or cost management, and quality management. These are vital strands in the management of both the corporate organisation and the project-based organisation. All sub-systems are governed by the organisational structural and operating parameters set by the internal and external influences. A sub-system will have management elements in common with other sub-systems and will also have elements unique to itself. For example, all sub-systems should be guided by a clear statement of organisational policy, while environmental management would in addition have the unique management element of assessing the environmental effects of the business and its operations.

The management elements in each sub-system provide structure and organisation to the human resources which support that sub-system. To enact these elements, sets of procedures and tasks at both the corporate level and project level are established. For this reason, if no other, management systems can best be described as organisational protocols and sets of procedures which organisational personnel implement in accomplishing their roles, as suggested previously in this chapter.

In many organisations, and construction organisations in particular, two distinct tiers of management will be developed within each sub-system. One will deliver specific management functions throughout the corporate organisation, while the second will deliver specific functions at the construction project level. Essentially, the macro sub-system is established at the organisation's head office, and within this framework a micro sub-system is established

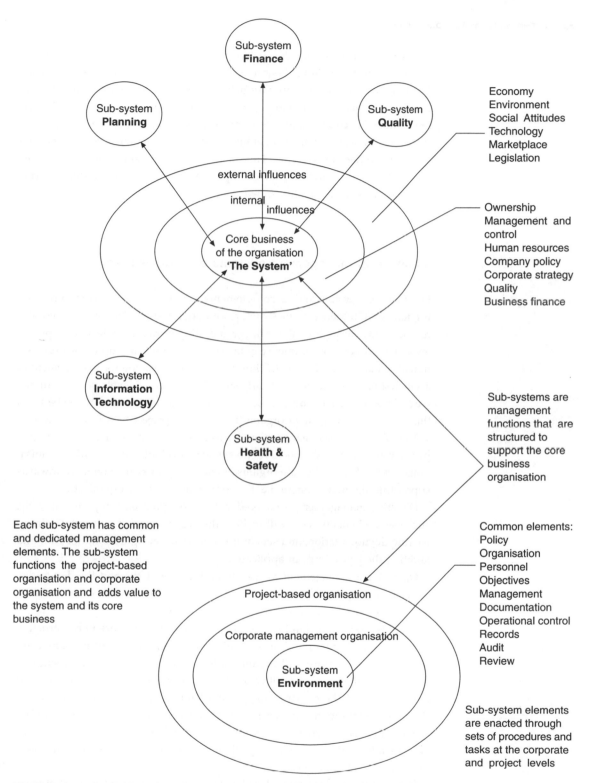

Sub-system **Finance**

Sub-system **Planning**

Sub-system **Quality**

Economy
Environment
Social Attitudes
Technology
Marketplace
Legislation

external influences

internal influences

Core business
of the organisation
'The System'

Ownership
Management and
control
Human resources
Company policy
Corporate strategy
Quality
Business finance

Sub-system
**Information
Technology**

Sub-system
**Health &
Safety**

Sub-systems are
management
functions that are
structured to
support the core
business
organisation

Each sub-system has common
and dedicated management
elements. The sub-system
functions the project-based
organisation and corporate
organisation and adds value to
the system and its core
business

Common elements:
Policy
Organisation
Personnel
Objectives
Management
Documentation
Operational control
Records
Audit
Review

Project-based organisation

Corporate management organisation

Sub-system
Environment

Sub-system elements
are enacted through
sets of procedures and
tasks at the corporate
and project levels

Fig. 8.1 An organisation viewed as sub-system elements supporting the system and its core business

343

for each site. The relationship between the macro and micro sub-systems is significant. Systems and sub-systems must be interactive in the widest sense. There must be close compatibility between the two, as each is mutually supportive and must link in and support the core business system. In addition, there must be commonality and consistency in approach within the sub-system, as each should have an ability to share information across the boundaries with other sub-systems. This might involve passing information from one project site to another, for example, or the finance sub-system sharing information with the quality sub-system.

Influence of management systems on organisation

Establishing a system and its complementary sub-systems in a company has a fundamental influence on its structure and organisation. As a system allocates responsibility throughout the entire organisation from corporate level to project level, the most obvious influence of the systems approach is upon the organisation of management. Figure 8.2 illustrates the organisation for management of a sub-system. Management is clearly stratified into broad levels: srategic, directive and operational. In addition, sub-systems implementation is shared between the corporate management organisation and the project-based organisation.

The role of strategic management is vested in executive management, whose primary function is to sanction sub-system development, provide organisational leadership, and guide organisational and project management towards supporting the main system, i.e. the core business of the organisation.

Directive management is charged with developing and implementing the sub-system elements. This will include the organisation of personnel, administering documentation, and sets of procedures covering training for staff and auditing the procedures in application.

Operational management focuses upon the project site and the day-to-day activities of supervisors, who operate or implement the sets of procedures, and who monitor and report back on their performance.

Organisation of this kind is typical of sub-systems structure in construction industry contracting organisations. This can be seen in practice in many companies in the structure and organisation of, for example, quality assurance, environmental impact, or health and safety management. The organisation and management of the construction process lends itself to a systems approach through specialisation, which is intrinsic to the construction processes (Griffith, 1997). Large and complex projects are characterised by their teams of specialists, which traditionally have included planners, estimators, surveyors, engineers and construction managers.

In recent times, these specialists have been joined by a complement of managers serving more recently accepted concepts such as quality assurance, health

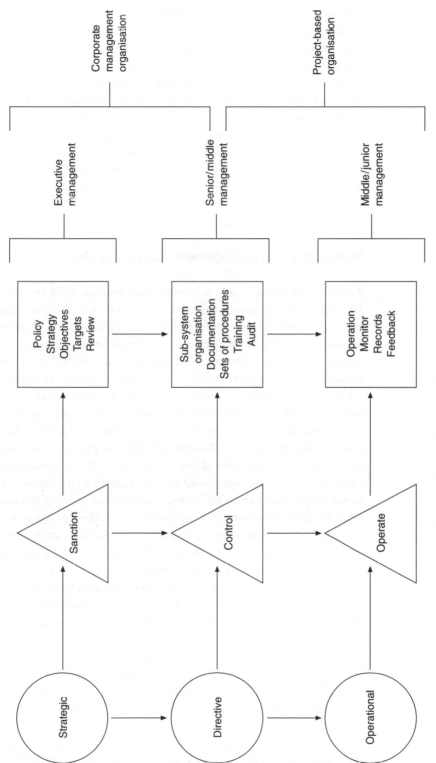

Fig. 8.2 Management stratification within the organisation of a sub-system

and safety, and environmental impact. Construction simply could not operate without such specialisation. Each specialist contributes through implementing particular sets of procedures reinforced by long-established professional working practices. Specialists guide the various stages of the total construction process, and the management functions, or sub-systems, they implement help them to execute the many, varied and complex arrangements that need to be made to make the construction process work in practice. Furthermore, today's construction clients and the corporate management of construction organisations require an assurance that the company and project procedures are being followed. This is particularly important where independent monitoring or auditing is used to determine organisational performance.

Sub-systems as management support services

The many sub-systems used in construction processes must be seen in context. Each represents a *support service*, that is, a service which can be paramount to project team effectiveness but which is not itself a producer of direct financial income. Support services augment both the corporate organisation and each temporary project organisation by providing a service. They interact with, influence and are a prerequisite to maintaining the core business by linking into the company's parent system. To be cost-effective to an organisation, each sub-system must be efficient and effective in serving both the project organisation and the corporate organisation. It is often unfortunate that the same specialisation that can contribute so effectively to project success can so easily create difficulties, typically at the boundaries between specialists. As additional sub-systems are introduced into the construction process to meet more demanding requirements, the boundaries between sub-systems can be indistinct, resulting in procedures which can become ethereal. When this happens, the intended benefits of the management sub-system can be lost, affecting both the cost-effectiveness of the individual project and the well-being of the corporate organisation. As poor performance becomes a reality on one construction project, so pressure is placed on other projects for greater success in compensation. From a corporate perspective, such failure invariably means that organisational overheads must be redistributed across a narrower spread of business, so reducing profit margins.

The best way to ensure that management sub-systems are both efficient and effective is to ensure that all personnel understand clearly the role that the sub-systems are fulfilling. The sub-systems exist to structure and organise management and working procedures at both corporate and project levels, and these procedures exist to serve the organisation's core business.

Such an approach requires an astute ethos and culture to be developed in the organisation. A systems philosophy requires the organisation to:

- adopt a holistic, or whole-organisation, culture in which management drive and commitment comes from the top and is actively encouraged to permeate the entire organisation;
- formally structure its management system and sub-systems to develop uniform procedures which support activities across the various levels of management;
- present a well-defined framework to organise and deploy effectively the resources necessary to support the management sub-systems; and
- be proactive in developing the necessary dynamic characteristics to ensure that sub-systems interact effectively with others and deliver the requisite service at both project and corporate levels.

Influence of management systems on human resources

Many companies in the construction industry have, in recent years, tended to rationalise and simplify their organisational structures. Management activities which have proved to be less than vital to the core business have been restructured or removed, and many non-essential functions have been shed, with the resulting loss of jobs. Contracting organisations, for example, are competing fiercely on costs with little if any margin for profit making, are niche marketing their services more than ever before, and must respond more flexibly to increasingly demanding client requirements. There has also been a tendency towards subdividing the business into independently managed strategic operations often founded on functional specialisation, where management has fully devolved responsibility for business performance. They are accountable to but are directly supported less by the corporate organisation.

Many traditional contracting organisations have become more adept in subcontracting. This effectively reduces their direct responsibility for people management within the project organisation. However, it does not eliminate the management function, as their subcontractors must be effectively co-ordinated and managed.

Certainly, the changing nature of work in construction processes has promoted leaner organisations with a greater degree of specialisation. These characteristics have tended to lend themselves to a systems management approach to organisation and operation. With this, there has been a refocus upon the ways in which human resources are regarded and valued by the more morphogenic organisations in construction.

Management and workforce groups tend to be organised into educated, informed and creative core elements, which the organisation regards as assets. Managers of such groups are, more than ever before, empowered to take responsibility for organising, motivating and leading these resources in managerially devolved activities. Effectively, they run their own business units within overall business parameters set by the parent organisation. Such a focus on human resource activity has led to stronger corporate and project foci for implementing structured support services for employees, for example 'Investors in People' (IIP). They also have a clearer direction for business activity to be more performance-led and output-measured. The added-value dimension of delivering customer satisfaction is uppermost, and both the organisation of the business and the activities of human resources are directed to this purpose.

The management systems described in this book demand that organisations do focus on specific activities to underpin the contribution of human resources. Management sub-systems can only be efficient and effective where, for example, clear communication, responsibility and feedback mechanisms are in place. In addition, because systems often introduce new procedures, working practices may need to be redefined and appropriate training given to both management and employees. This is why training is always included in any management system development. Organisational commitment to training can alleviate many of the problems associated with handling change. Personnel may fear new procedures and the perceived intent behind them, and educating and training systems managers and operatives will help to clarify misconceptions, allay potential fears and avoid organisational dysfunctionality. Where a management system is successfully introduced, the real investment by the organisation is not so much in developing the policies, organisation and procedures but rather in developing the human resources that implement and run the system.

Commitment to management systems

A classic dilemma of a systems approach is that as attention is given to a particular sub-system or specific elements within a sub-system, the integrity and efficiency of the holistic organisational system is often compromised. In such circumstances, many of the intended benefits of individual sub-systems may be lost as each sub-system competes for appropriate attention, adequate resources and management commitment.

A prerequisite to developing any organisational system or management sub-system is that it should achieve a gain in synergy for the corporate organisation in pursuit of its core business; that is, the whole must be greater than merely the sum of the parts. The various management sub-systems must seek to establish synergistic linkages with other sub-systems and the organisation's parent system. At the same time, the sub-systems must also recognise their own distinct and often necessary differences. These differences will, for example, respect particular legislation or the formal system specifications of industry sector accreditation bodies.

Unfortunately, achieving optimum synergistic benefit is far from straightforward. A number of key reasons predominate. First, the tradition with construction for specialisation means that the structural separation of management functions can be counterproductive to effective resource allocation. Second, the divide between the two tiers of management within a sub-system, i.e. the corporate tier and the project tier, can make sub-system interaction less than effective. Third, where there are a number of independent management sub-systems in operation, there will always be some level of competitive demand for resources or priority attention. Given these characteristics, which frequently occur in construction, synergy can clearly be difficult to achieve.

Synergy within the system can only have a chance of success if the utmost commitment is given by the organisation. There are a number of prerequisites to management systems development. These apply at corporate and project level and focus on management commitment and planning. Development vision and support for the system and its sub-systems must be demonstrated by corporate management and encouraged to develop holistically within the wider organisational context.

To ensure that each sub-system has an optimum chance of being effective, there needs to be:

- a demonstrable commitment to organisational policy and strategy in the specific conceptual management area, e.g. quality, IT, health and safety.
- a clear statement of organisational ethos and policy, which should be circulated throughout the organisation.
- employee ownership of the sub-system management function through involvement in formulation, development and implementation.
- identified goals and targets at both corporate and project levels against which group performance can be measured.
- adequate resources to facilitate the sub-system framework and operational working.
- appropriate and continuing education and training for management and the workforce in operational procedures.
- on-going review and improvement of applications to enhance employee experience and expertise.

Similarly, at project level, prerequisites to effective sub-system utilisation include the following:

- identification of key issues that need to be addressed on the project site within the management function sub-system.
- risk assessment within the specific management function.
- development of action plans in response to identified needs.
- distribution of good practice guidelines to functional staff.
- determination of audit procedures between the corporate system and the project sub-systems.

- briefing of all project team members on sub-system operation.
- training in the use of procedures that will be used to plan, monitor and control the management function.
- practice notes on good management procedures to be adopted.
- item checklists for the monitoring of specific procedures.
- self-audit and review documents for the activities of sub-system supervisors.
- guidance notes on potential actions when problems occur.
- references to corporate management and other sub-systems where assistance may be needed.

These prerequisites to the development of management sub-systems will have been seen clearly in some of the earlier chapters describing their applications to specific concepts such as health and safety and environmental management. It is absolutely essential that an organisation carefully studies its sub-systems and the deployment of human resources. Both management and workforce need systems that are efficient, effective and clear and easy to understand and implement if benefits are to be achieved and organisational synergy is to be maximised.

References

Anderson, A. and Marshall, V. (1996) *Core Versus Occupation Specific Skills*, Department for Education and Employment (DfEE), HMSO, London.

Griffith, A. (1994) *Environmental Management in Construction*, Macmillan, Basingstoke.

Griffith, A. (1997) *Towards an Integrated System for Managing Project Quality, Safety and Environmental Impact*, Australian Institute of Building (AIB) Papers, Vol. 8, pp. 67–77, Melbourne.

Griffith, A. and Watson, P. (1998) *Optimising Management Systems for Construction*, Australian Institute of Building (AIB) Papers, Vol. 9, Melbourne.

Index

accident rates 4, 149
accreditation 105
activities 25
activity box configuration 60
activity duration 23, 27
Acts of Parliament 152
advanced software tools 313
allocating resources 57
applications packages 303, 312, 332
appointment of parties 149
arrodence diagram 30–1, 62
arrow diagram 22–9, 38
as-built networks 40
assessment 105
authentication 326, 328
average annual percentage rate of
    return 88, 95

backbone network 318
backward pass 25, 27, 29
bar charts 18–20, 62
basic network principles 23
Blackspot Construction 149
bridges 320
briefing process 283–5
BS 7750 6, 253–4
BS 8800 150
BS EN ISO 9000 104, 105, 108, 109,
    113, 115, 117, 119, 124, 141,
    142, 147
BS EN ISO 9001 108, 128
BS EN ISO 9002 108
BS EN ISO 9003 108
budgetary control 76, 84, 87
budgets 76
business strategy 307

calendars 52
Capital Investment Appraisal 88
cash flows 75, 76, 81
causality 123
centralised computing 317
certification 264–5
change 340
claims 39

client/server computing 305, 321–3,
    337
clients 4, 161, 254
communication 300–1, 315–17
    channel 318
    of information 315, 324–5, 337
competition 340
completion dates 37
complex compression 45, 62
complex interfacing 34–5
complex network 31–3
compression 45–8, 62
computer applications software 11,
    51
connectivity 315
constraints 39, 55–6
Construction (Design and
    Management) Regulations
    1994 155, 157–9
Construction-Phase Health and
    Safety Plan 168, 176, 206
contract
    planning 14–15
    programme 13, 16
    stage 15
contractors 4, 164
control 75
controlling progress 37, 48, 62
core business 342
corporate organisation 344
cosmic view 59, 61
cost curves 45
cost slopes 43, 46–7
costs and benefits of IT 308–9, 336
cost-significant items 13
CPM definitions 27
CPM equations 27
crash cost 42–3, 46
crash duration 43, 46
crash time 42
crashing activities 42
critical activities 40, 46
critical path 27, 35, 37, 42, 45, 48
critical path method (CPM) 23
cummulative costs 334

daily work operations 17
data back-up 335
data capture 303
Decision Making Process 64
decompression 45–6, 62
defining activities 22
defining calendars 54
defining resources 57
demodulation 316
dependencies 32
design 284
    detailed design 288–9
    scheme design 285, 286–8
designers 162, 282
direct changeover 330
directive management 344
disaster recovery plan 328
disbenefits of IT 309
discrete costs 44
disputes 39
distributed environment 327
distributed processing 317
distributed systems 332
documentation 167, 207, 334
dummy activities 26, 29
duration of activities 23

effective information systems
    management 333–5
electronic bulletin boards 317
electronic data interchange (EDI)
    317
electronic mail 317
electronic trading 305
employees 164
encryption 326, 328
end-user computing 305, 313, 332
environmental awareness 256
environmental considerations 258
environmental effects 255–8
environmental impact assessment
    (EIA) 279–82
environmental management system
    (EMS) 265
    audit 277–8

checklists for system development 294–7
documentation 275
effects 272–3
objectives 273–4
operational control 275–7
organisation 271–2
policy 269–71
programme 274
records 277
review 278–9
estimated allowances 17, 24, 40, 53
European Directives 153
European Eco-Management and Audit Scheme (EMAS) 254
evaluation 329
events 25
extranets 324–5, 337

Factories Act 1961 152
firewalls 327
Fixed and Variable Costs 65
float 27, 29
forward pass 25, 27, 29
fragnets 48
future developments 335–6

gateways 320
group support systems (GSS) 303, 325–6, 337
groupware 303, 325–6, 337

hardware and software selection 329–30, 337
Health and Safety
  at Work, etc Act 1974 152
  Commission 149
  common law 151
  Executive 161
  File 168–9
  legal system 151
  Plan 5, 168, 173–9
  statute law 152
heterarchy 122
holism 340–1
human resources management 347–8
human-computer considerations 331–2
hybrid network 317

implementation 337
implementation methods 330–1
implementing information systems 329
indirect costs 42
information 300–1, 315

information management 305, 337
information systems 305, 310
intangible benefits 308
integrated services digital network (ISDN) 320
integration 303, 308, 315
intelligent components 336
internal rate of return (IRR) 89, 97, 219
internet 323–4, 326
internet service provider 323
interrelationships 20
intranets 324–5, 337
ISO 14001 253

ladder contructs 33
lag time 33
lead time 33
least cost 45
linear relationship 43
liquidated and ascertained damages 12
local area networks (LANs) 317–21, 325, 330
logic diagram 38
logic relationships 29

maintenance contracts 335
management
  commitment 348–50
  concepts 341
  considerations 334–5
  functions 342
  of health and safety 169
  systems 341
Management of Health and Safety at Work Regulations 1992 153
managing resources 11, 40–2
margin of safety 67
Marginal Costing and Break-even Analysis 64, 70, 73
material resources 51
measured work 13
methods of analysis and design 310
methods of development 337
methods of programming 17–18, 62
metropolitan area networks (MANs) 317–18
mixed-based (semi-variable) costs 65
mobile networks 336
modems 303, 320
modernist theory 120
modulation 316
monitoring 106
monitoring progress 37–40, 51, 62
morphogenic 123
morphogenic organisation 341

morphostatic 123
morphostatic organisation 341
multimedia 303, 323
multi-project planning 47–51

negative float 37
net present value (NPV) 309
network
  analysis 18–20, 62
  installations 327
  logic 25, 52
  non-complex 32
  privacy and security 327–8
networks 305, 315, 332
non-critical activities 18, 40, 46, 55
non-linear relationship 44
normal cost 43
normal duration 43

operational management 344
optimum cost 46
optimum project duration 45
organisation 340
organisational culture 325
outsourcing 313–14, 328
overall programme 13, 21

parallel systems 330
payback assessment 309
payback method 88, 92
percentage completion 37
PERT view 59
phased-in systems 331
pilot systems 330
planner's role 13
planning and control 11–13
Planning Authority and building control 161
planning software 51–2
Planning Supervisor 162, 165–6
post implementation review 333–4
post-modernist 120, 121
precedence diagram 22, 28–30, 39
pre-contract planning 14–15
preparatory environmental review 268–9
pre-qualification 264
Pre-Tender Health and Safety Plan 168, 173–6
pre-tender planning 13–14
Principal Contractor 5, 163, 166–7
  health & safety management system (H&SMS) 185
  health and safety management policies 187–8
  permits to work 189–90

project administration: site rules
and management procedures
191
risk assessment 169–73, 188
safety induction and training
190–1
safety management audits 192–3
safety method statements 188–9
system examples 193
system pro-formas 193
private intranets 325
private networks 319
procedures manual 109, 110
profit graph 68
profitability 12
progressing work 11, 37–9, 61–2
project
environmental familiarisation
292–3
environmental management
292–3
environmental planning 291–3
organisation 344
overheads 12
prototyping 312
public networks 319
public switched data network
(PSDN) 320
public switched telecommunications
network (PSTN) 319

Quality Assurance Systems 103, 129
quality manual 109, 110
quality policy 128

real-time
communications 303
dummies 33
recovery procedures 335
relationships 29, 52, 56
resource
aggregation 40
allocation 40, 42, 53, 55

management 62
profile 41, 58
requirements 13, 48
smoothing 40–1, 55
table 58
restraints 32
return on investment (ROI) 309
risk assessment 169–73, 188, 290
evaluation of risk 170
hazard identification 170
prevention and protection
measures 172
routers 320, 327

Small Company Environmental
and Energy Management
Assistance Scheme
(SCEEMAS) 265
sectionalised network 33–5
sensitivity analysis 309
short-term planning 13, 17–19
simple interfacing 34
software
considerations 51
evaluation and selection 331
upgrades 335
specialisation 344
specification for systems 261
staff education and training
332–3
stages of planning 14
standard 's' curve 80
standards 263
strategic issues of IT 306–7
strategic management 344
subcontractor work 23, 51
sub-networks 34–6
sub-systems 342–4
suppliers 23
support services 342
SureTrak Project Manager 2.0
52–62
switched networks 318

system 341–4
development considerations
314–15
implementation 307, 332
support 335

tangible benefits 308
target bars 59
target dates 14, 59
technical manuals 335
telecommunications 303, 316–17,
327, 336–7
tendering 16
time/cost optimisation 42–7, 62
time-based costs 65
total costs 66
Total Quality Management 141, 142,
143, 144, 145
traditional system development life
cycle 310–11
training 51
training in IT 332–3
transmission 316
types of programme 15–17

UK legislation 153
updating activities 60
upgrading 329
user documentation 51

value-added networks 319
video-conferencing 303, 317
virtual private networks 326–7, 337
voice mail 317

web browsers 324
web sites 325
wide area networks (WANs) 317–19,
325
wireless networks 336
work breakdown structure 22
work durations 24
world wide web (WWW) 323–4, 337